Environmental Footprints and Eco-design of Products and Processes

Series Editor

Subramanian Senthilkannan Muthu, Head of Sustainability - SgT Group and API, Hong Kong, Kowloon, Hong Kong

Indexed by Scopus

This series aims to broadly cover all the aspects related to environmental assessment of products, development of environmental and ecological indicators and eco-design of various products and processes. Below are the areas fall under the aims and scope of this series, but not limited to: Environmental Life Cycle Assessment; Social Life Cycle Assessment; Organizational and Product Carbon Footprints; Ecological, Energy and Water Footprints; Life cycle costing; Environmental and sustainable indicators; Environmental impact assessment methods and tools; Eco-design (sustainable design) aspects and tools; Biodegradation studies; Recycling; Solid waste management; Environmental and social audits; Green Purchasing and tools; Product environmental footprints; Environmental management standards and regulations; Eco-labels; Green Claims and green washing; Assessment of sustainability aspects.

More information about this series at https://link.springer.com/bookseries/13340

Subramanian Senthilkannan Muthu
Editor

Toys and Sustainability

Editor
Subramanian Senthilkannan Muthu
Head of Sustainability - SgT Group and API
Hong Kong, Kowloon, Hong Kong

ISSN 2345-7651 ISSN 2345-766X (electronic)
Environmental Footprints and Eco-design of Products and Processes
ISBN 978-981-16-9675-6 ISBN 978-981-16-9673-2 (eBook)
https://doi.org/10.1007/978-981-16-9673-2

This Springer imprint is published by the registered company Springer Nature Singapore Pte Ltd.
The registered company address is: 152 Beach Road, #21-01/04 Gateway East, Singapore 189721, Singapore

Contents

About the Editor

Dr. Subramanian Senthilkannan Muthu currently works for SgT Group as Head of Sustainability and is based out of Hong Kong. He earned his Ph.D. from the Hong Kong Polytechnic University and is a renowned expert in the areas of Environmental Sustainability in Textiles & Clothing Supply Chain, Product Life Cycle Assessment (LCA) and Product Carbon Footprint Assessment (PCF) in various industrial sectors. He has five years of industrial experience in textile manufacturing, research and development and textile testing and over a decade of experience in life cycle assessment (LCA), carbon and ecological footprints assessment of various consumer products. He has published more than 100 research publications, written numerous book chapters and authored/edited over 100 books in the areas of Carbon Footprint, Recycling, Environmental Assessment and Environmental Sustainability.

List of Figures

Design and Production Process of Toy Prototypes Using Urban Forestry Waste

Tips for Selecting Wood from Urban Afforestation for the Production of Toys: How the Sustainable Reuse of Waste Can Result in Economic, Environmental and Social Benefits

Design and Production Process of Toy Prototypes Using Urban Forestry Waste

Luiz Fernando Pereira Bispo, Adriana Maria Nolasco, Allana Katiussya Silva Pereira, João Gilberto Meza Ucella-Filho, Fabíola Martins Delatorre, Gabriela Fontes Mayrinck Cupertino, Regina Maria Gomes, Elias Costa de Souza, José Otávio Brito, and Ananias Francisco Dias Júnior

Abstract Urban wood wastes of some species have mechanical, physical, and chemical characteristics that classify them as suitable for use in the production of small wooden objects, such as toys. Wooden toys market moves a significant amount of money in several countries around the world, and it is important to know more about possible alternatives that can complement this sector. Thus, the objective of this book chapter was to describe the creative design and the production process of toy prototypes using wood wastes from urban forestry. To this end, we made sketches of the products based on the study of the desired design for the final production of the toys. From this, we discussed the entire process of production of toys from wood wastes from urban forestry, which goes from cutting trees and branches, through the process of transport to industry, selection of suitable wooden parts for production, drying, and storage of wood, mechanical processing of wood, painting of toys, to the manufacture of toys for sale. We discuss the main aspects of each stage of the production process, evaluating their greatest difficulties and strategies necessary for good performance in each one of them, which will result in the production of safe and good quality wooden toys for consumers. The results found can serve as a basis for future projects that assess the economic feasibility of implementing the production of wooden toys from the use of wood wastes from urban forestry.

Keywords Urban wood waste · Sustainable production · Sketching · Toy designer · Wooden toys · Wood machining

---

L. F. P. Bispo · A. M. Nolasco · R. M. Gomes · E. C. de Souza (✉) · J. O. Brito
"Luiz de Queiroz" College of Agriculture, University of Sao Paulo, Piracicaba, SP, Brazil

A. K. S. Pereira
State University of Southwestern Bahia—UESB, Candeias, BA, Brazil

J. G. M. Ucella-Filho · F. M. Delatorre · G. F. M. Cupertino · A. F. Dias Júnior
Department of Forestry and Wood Sciences, Federal University of Espírito Santo, Jerônimo Monteiro, ES, Brazil

S. S. Muthu (eds.), *Toys and Sustainability*, Environmental Footprints and Eco-design of Products and Processes,
https://doi.org/10.1007/978-981-16-9673-2_1

1 Introduction

When the environmental challenge at hand is the issue of waste, different civil society groups have drawn the attention of decision makers and public opinion to a particular villain: plastic. In search of modernity, the arrival of the thermoplastic resin industry in the world made it possible to reduce production costs, especially of manufactured goods for domestic and personal consumption, such as shoes, toys, and packaging. Pieces that used to be made of materials such as wood, glass, and paper began to have a more accessible, modern, practical, and cheap aesthetic (Zamora et al., 2020), but not very sustainable from an environmental point of view.

Global plastics production totaled 368 million tons in 2019 (Tiseo, 2021). It is estimated that production in 2020 decreased by about 0.3% due to the impacts of COVID-19 on the industry, but the versatility of this group of materials means that there is continuous growth in its production year after year (Tiseo, 2021). As a result, the environmental impacts generated by plastic products discarded as garbage have been growing exponentially (Manzoor et al., 2020; Okunola et al., 2019), causing the reuse or recycling of this material (Prata et al., 2019; Seghiri et al., 2017), and the search for alternative products becomes increasingly necessary.

One of the alternatives for the development of polyethylene-free objects is the use of lignocellulosic materials. Companies such as "Melissa & Doug," "Friendly Toys," and "Kind to Kidz" are examples of brands that sell toys made from wood, a lignocellulosic material that can be supplied, for example, by urban forestry waste. The production of this waste is common in all cities in the world, and in the United States alone, an annual generation of around 33 million tons of urban dry wood is estimated (Nowak et al., 2019). Based on this volume of material produced, waste from urban trees can be used to produce multiple products (de Meira et al., 2021; Nasser et al., 2016; Nowak et al., 2019; Palharini et al., 2018) as a result, mainly, of its availability and heterogeneity, also serving as a potential source of raw material for the manufacture of unconventional wooden toys (Bispo et al., 2021) to replace those made with plastic.

The toy manufacturing process using urban wood waste involves crucial steps ranging from the cutting of trees and branches, through the selection of wood parts suitable for the production, drying, storage, and technological characterization of wood, to the manufacture of toys for sale. All of this is done from the creation of prototypes governed by planning that aims to establish the structural, aesthetic-formal and functional characteristics of the object (Bonsiepe, 1997), considering essential aspects such as which stage of the child's development the product is intended for.

The manufacture of toys from waste wood allows the development of a more ecological product, as it allows the use of fewer inputs. When compared to conventional wooden toys already found on the market, toys from urban wood waste have advantages, mainly because they require fewer operations and, therefore, less processing of the raw material. This is because the application of this material may involve the maintenance of the natural shape of the bark, dimensions, contours of branches (De Meira, 2010) and trunks, resulting in a reduction in operating costs.

Toys produced with waste from urban trees can also become a source of income for small entrepreneurs. Bispo et al. (2021) indicate that these objects can cost, based on the conception of potential consumers, around $3.70 and $13.58, with a lower value attributed to more rustic toys and higher values to those that have undergone machining processes. In addition, the use of this waste applied to create wooden objects, especially children's toys, contemplates four of the seventeen sustainable development goals (SDGs) established by the United Nations (UN), namely: reducing inequality; responsible consumption and production; partnerships and means of implementation; sustainable cities and communities (United Nations, 2021).

Based on the context presented, this chapter aims to describe the creative design and the process of producing toy prototypes from wood waste from urban afforestation, in order to facilitate their creation and, consequently, contribute to sustainable and socioeconomic development.

2 The Role of Toys and Games in Children's Development

The first years of life correspond to the period of greatest importance for children in terms of sensory stimuli, curiosity about the world, and the development of human personality and identity. From a very early age, children maintain learning and developmental relationships through interactions with other individuals that grow stronger over time and significantly influence the maturation process. The act of playing takes place in order to provide children with a world of their own (Vygotski, 1991), showing society how recreational activities are fundamental instruments in any and all early childhood education institutions (Kishimoto, 1995; Vygotski, 1991).

Games and toys arise from adult practices, which involve religious and astrological rituals, and representations of nature. Archaeological records show that wooden toys have been produced since the Stone Age (Cywa & Wacnik, 2020). Nowadays, there are toys made of multiple raw materials, found all over the world, mainly among the American Indians, African blacks, primitive people from Australia and Polynesia, indigenous people from Asia, and Brazil (Kishimoto, 2014), and they are considered to be an opportunity for caregivers to support children's growth (Goldstein, 2012; Healey & Mendelsohn, 2019).

The role of play in learning has always been considered of primary importance. Thus, programs aimed at early childhood are composed of attributes that offer multiple possibilities for play and interaction between children, ideas, and other people (Yelland, 1999). Based on this, one of the fundamental premises of early childhood education is the belief that children learn better through games (Vygotski, 1991; Yelland, 1999). Such an assumption is based mainly on classical, ideological, philosophical and pedagogical principles such as Montessori, Steiner, and Froebel (Yelland, 1999). Thus, toys are considered essential in children's early childhood development, especially in relation to their facilitation of cognitive development, language interactions, symbolic and pretend play, problem solving, social interactions, and physical activity, with increasing importance as children mature (Goldstein, 2012).

In the context of learning, playful behavior promotes positive results, such as greater pleasure (Rubens et al., 2020), involvement (Ofer et al., 2018; Yu et al., 2020), and increased motivation (Liu et al., 2019), thus contributing to healthy child development (Lee-Cultura et al., 2021). By playing, children learn to make decisions and understand the world better. Through playful toys and games, important aspects of their development are developed, for example, creativity and the ability to make decisions (Healey & Mendelsohn, 2019; Vygotski, 1991), and cognitive, physical, social, and emotional well-being (Healey & Mendelsohn, 2019).

Within these dimensions of play, toys play an important role in triggering our imagination and motivation, laying the groundwork for improving cognitive and motor skills, in addition to teaching us the importance of sharing, cooperating, and communicating (TIE—Toy Industries of Europe, 2021). In this context, it is possible to note the importance of coexistence with and use of toys, as it is from them that the child learns to act in a cognitive, imaginative sphere, rather than in a purely visual sphere, only with what they can see and touch (Vygotski, 1991).

Technological advances, modernization, and high connectivity have meant that children are exposed to greater innovation than previous generations, resulting in greater contact with interactive toys that have technology as the basis for providing games (Sridhar et al., 2017). However, traditional toys can be more interesting for a child to venture out and explore, especially if the child watches adults handling them.

Many caregivers believe that high-cost, screen-based electronic toys such as tablets are essential for their children's healthy development (Levin & Rosenquest, 2001). However, evidence suggests that the core elements of such toys (e.g., lights and sounds emanating from a robot) detract from the social engagement that could occur through toys and games that allow the use of facial expressions, gestures, and vocalizations (Klin et al., 2003), like traditional toys, which include wooden toys.

Plays involving traditional toys such as blocks and puzzles allow problem solving and, consequently, cognitive, language, fine motor skills development and encourage early mathematical and spatial skills (Ferrara et al., 2011; Weisberg et al., 2013). In addition, such traditional toys tend to provide greater involvement on the part of caregivers, facilitating imaginative play and problem solving in such a way that electronic toys are less likely to do (Zosh et al., 2015).

All these positive aspects that toys made with wood can provide are only possible when the manufacturing process involves fundamental aspects of design, the main one being the incorporation of structural, aesthetic-formal, and functional characteristics (Bonsiepe, 1997), considering the development phase of the child for which the toy is intended.

3 Description of the Toy Creation Process with Urban Forestry Waste

In the midst of so much technology, making toys from renewable materials is an art capable of guaranteeing very attractive aspects, such as sustainability, lower production costs, and ample possibilities for shapes and design. Urban forestry wastes have physical, chemical, and mechanical characteristics that classify them as potential for the production of toys. Thus, this topic presents in detail the step by step to be followed to facilitate the toy manufacturing process from the use of residual biomass from trees used in urban afforestation (Fig. 1).

The first step of the production process comprises the collection of this material. Urban forestry waste comes from pruning and removal of arboreal individuals located in public and private areas, also including several trees that are naturally felled due to bad weather, such as rain and lightning. Such material can be composed of twigs, branches, stems, and roots, as shown in Fig. 2.

Identifying the species of wood waste is an important step for the development of products with lignocellulosic materials. If it is a species with no studies on its suitability for manufacturing toys, the characterization of its properties is essential to infer about the quality and durability of the product to be generated. Therefore, it is recommended to apply analysis of macroscopic, organoleptic parameters (color, shine, odor, taste, grain, and texture), basic and apparent density, shrinkage, janka hardness, and chemical composition, especially in relation to the types of extractives, holocelluloses, and macromolecular composition of lignin, as suggested by Bispo et al. (2021).

These characterizations determine the potential application of wood waste for the development of toys. Thus, it is suggested that materials with higher density be used in the manufacture of objects that require greater impacts during their use. On the other hand, less dense woods are indicated for the development of board game

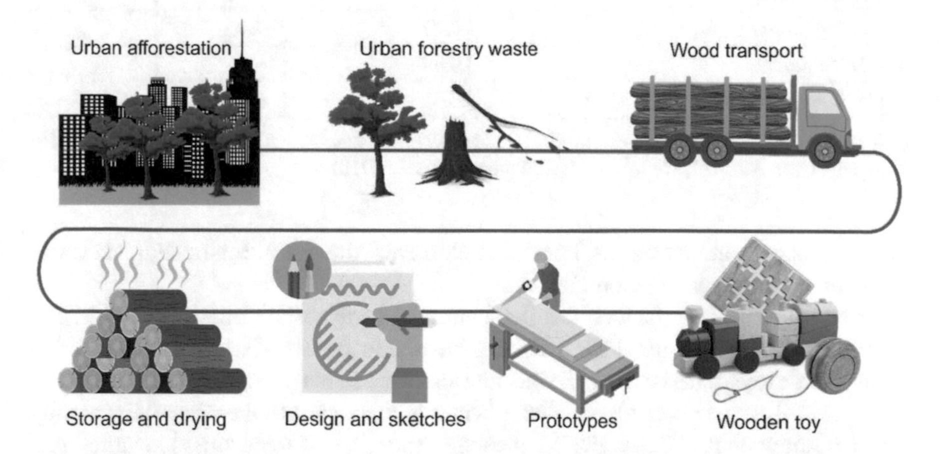

Fig. 1 Steps for creating toys with wood waste from urban afforestation

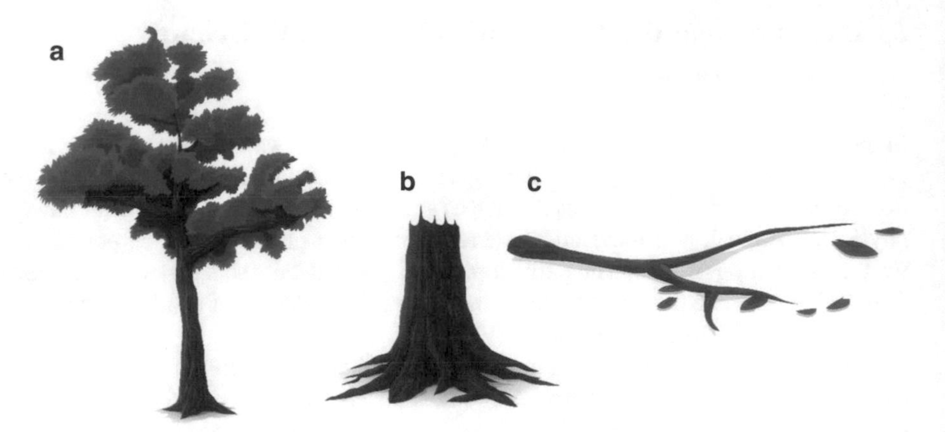

Fig. 2 Sustainable raw material for the manufacture of toys composed of urban wood waste. Where: (a) entire tree; (b) trunks and roots; (c) branches

Table 1 General characteristics of tree species used in urban afforestation with potential for the development of wooden toys. Where: BD = basic density; VS = volumetric shrinkage; TE = total extractives

Species	BD (g.cm^{-3})	VS (%)	TE (%)
Handroanthus heptaphyllus (Vell.) Mattos	0.70	11.38	6.33
Spathodea campanulata P. Beauv	0.30	12.20	7
Schinus molle L	0.47	12.49	7.01
Fagus sylvatica L	0.32–0.55	–	0.8
Caesalpinia peltophoroides Benth	0.68	11.01	–
Tibouchina granulosa (Desr.) Cogn	0.52–0.65	–	10.56
Lagerstroemia indica L	0.59	–	–
Terminalia catappa L	0.53	6.96	–
Ficus benjamina L	0.47	7.19	–
Nerium oleander L	0.44	–	–
Delonix regia (Bojer ex Hook.) Raf	0.44	6.81	–

(Arnstadt et al., 2016; Bispo et al., 2021; Ebner & Petutschnigg, 2007; Finžgar et al., 2014; Klingenberg et al., 2021; De Meira, 2010; Palharini et al., 2018)

pieces, puzzles, among others. For fitting elements, the use of lignocellulosic waste with lower shrinkage is recommended.

In addition to these factors, detailed chemical analyses are important to identify potentially toxic elements (PTEs) that can invalidate the use of wood for the creation of children's toys. This is due to the fact that some elements present in the raw material have the ability to become bioavailable when in contact with their handlers, mainly orally (Guney et al., 2020). Table 1 presents examples of trees already studied that provide material whose technological properties are indicated for the development of wooden toys.

After choosing the material, the design must be carefully planned. This is a crucial step in the manufacture of toys, as it is closely related to the consumer's perception (Faller, 2009). Design was defined by Bonsiepe (1997) as the design activity responsible for the structural, aesthetic-formal, and functional characteristics of a product for mass production. Thus, considering the manufacturing context, it is the designer's role to define which and how aspects will be met, based on the age group for which the toy is intended, such as attractiveness, level of interaction allowed between children and toys, and mainly, the characteristics of each material to be used.

The projection of a toy also requires that other aspects be considered. According to Viegas et al. (2014), the International Council for Children's Play defined criteria related to four fundamental qualities according to which a toy can be analyzed. The functional value concerns its adaptation to the user, taking into account its usability and safety. The experimental value, like toys that imitate professions, is related to what the child can do or learn with the toy, how to make noise, rotate, fit, build and measure. The structuring value allows for the assimilation of emotions and sensations and concerns the development of the child's personality. Finally, the relationship value is related to how the game or toy facilitates the establishment of relationships with other children and adults, proposing the learning of rules.

Risk situations that toys can involve children should also be considered by design. When the criteria for their use, age adequacy, and most importantly, the child's developmental stage are not respected, toys can become dangerous elements (Babich et al., 2014; Waksman & Harada, 2005). Thus, the requirements are differentiated depending on the age at which the toy is intended: children under 3 years old, 8 years old, or 12 years old. In the United States, toys come under the jurisdiction of the US Consumer Product Safety Commission (CPSC), which has issued guidelines to determine appropriate age groups for toys, and safety regulations are generally issued in accordance with the Consumer Product Safety Act (CPSA) and the Federal Hazardous Substances Act (FHSA) (Babich et al., 2014).

About the physical hazards associated with toys, they include suffocation with small parts, sharp edges, noise from toys that produce sound, and flammability. US and European standards include a range of tests to address physical and toxicological hazards. In the United States, manufacturers of certain children's products are required to certify that their products have been tested by a third-party laboratory and that they comply with all applicable standards (Babich et al., 2014). Accidents involve all social classes and all age groups, with specific characteristics for each of them (Waksman & Harada, 2005). The CPSC (2019) reported that in 2018 in the United States there were an estimated number of 226,100 toy-related injuries treated by the emergency department, with about 73% occurring in children under 15 years of age; 70% occurred with children 12 years of age or younger; and 37% percent happened to children under five years of age.

The manufacture of a toy involves a multidisciplinary team, and the design and processes must take into account all these aspects and then move on to the next step: the production of sketches. The sketch is a developed drawing, usually freehand, following the inventor's imagination based on the characteristics of the raw material and has the initial information necessary to execute the idea. Therefore, the sketch is covered with the emotion of the bodily manifestation, a stronghold, and product of the significant interaction between sensory, motor, and brain organs (Regal, 2003).

During the act of idealizing a toy with wood waste, the age of the target audience, attractiveness, functionality, and purpose of use must be taken into account, whether for educational purposes, such as board games and puzzles (Fig. 3), or for fun, such as carts, dolls, robots, trains, villages, among others (Figs. 4, 5, 6 and 7), always seeking to awaken the playful side of children. In addition, it is important to see the processing of the toy, so that when drawing the sketch one should look for designs that are easy to apply during manufacturing.

With the sketches in hand, selecting the raw material based on the proposed designs is a necessary step to facilitate the execution of ideas. When it comes to urban forestry waste, the collected biomass, especially the branches, has different dimensions, colors, and shapes. Thus, in order to take advantage of most of these materials, several toys can be generated based on their characteristics:

- Diameters smaller than 8 cm: ideal for small-sized objects with artisanal manufacturing, such as dice, puzzle pieces, checkers, chess, dolls, carts, geometric shapes, and letters.
- Diameters from 8 to 15 cm: provide the necessary dimensions for making robots, carts, animals, tops, yo-yos, geometric shapes, and tools (screws and drills).

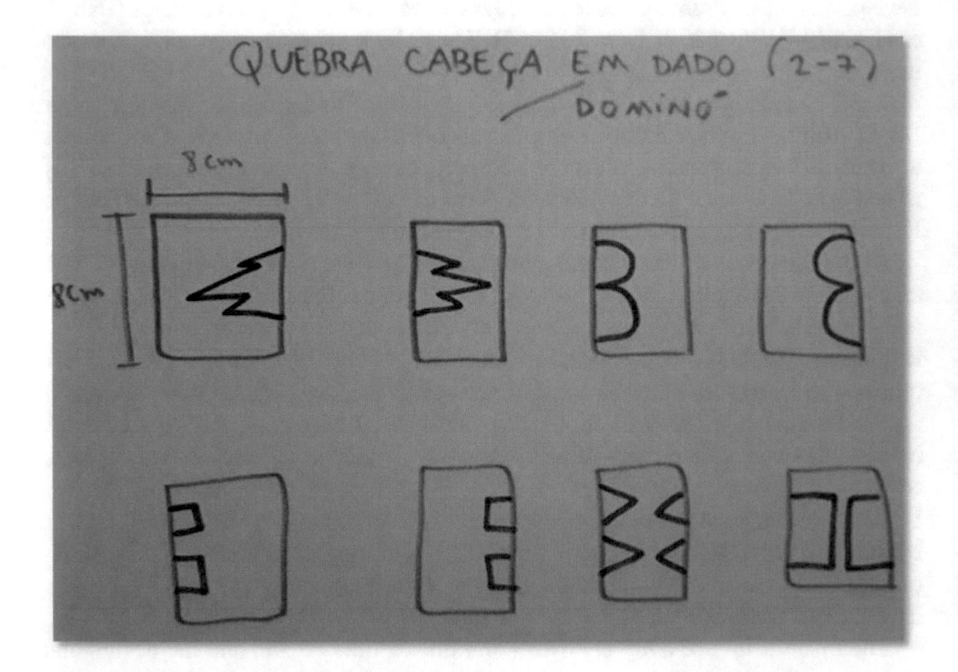

Fig. 3 Educational toy sketch (Puzzle) (*Source* Bispo et al., 2021)

Fig. 4 Sketches of village toys (*Source* Bispo et al., 2021)

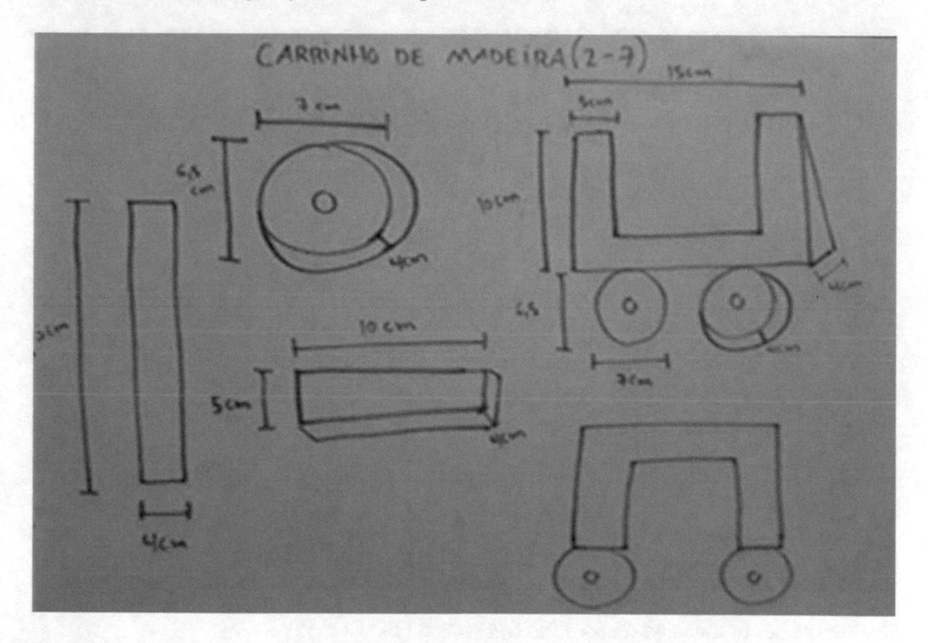

Fig. 5 Sketches of cart toys (*Source* Bispo et al., 2021)

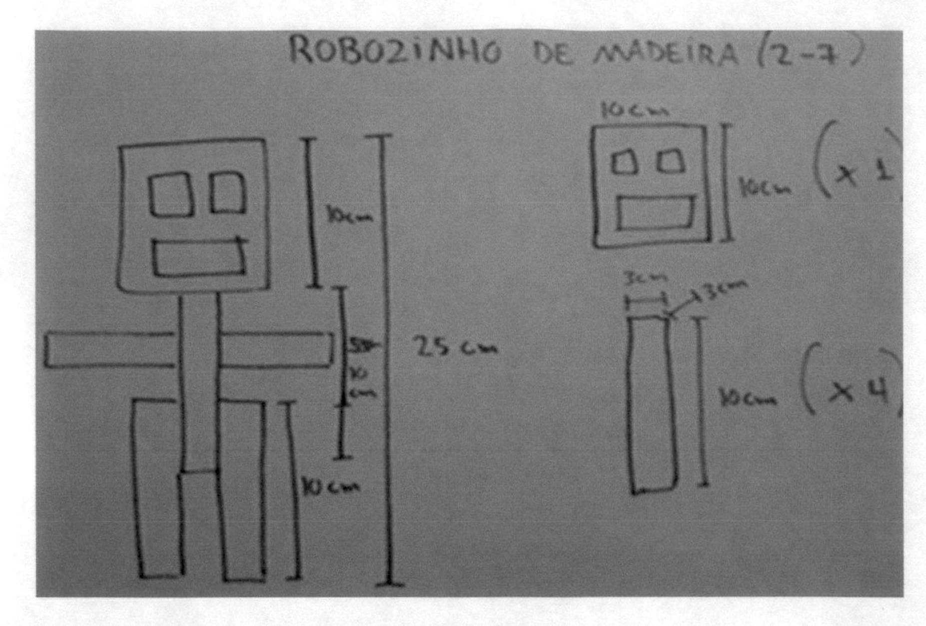

Fig. 6 Sketches of robot toys (*Source* Bispo et al., 2021)

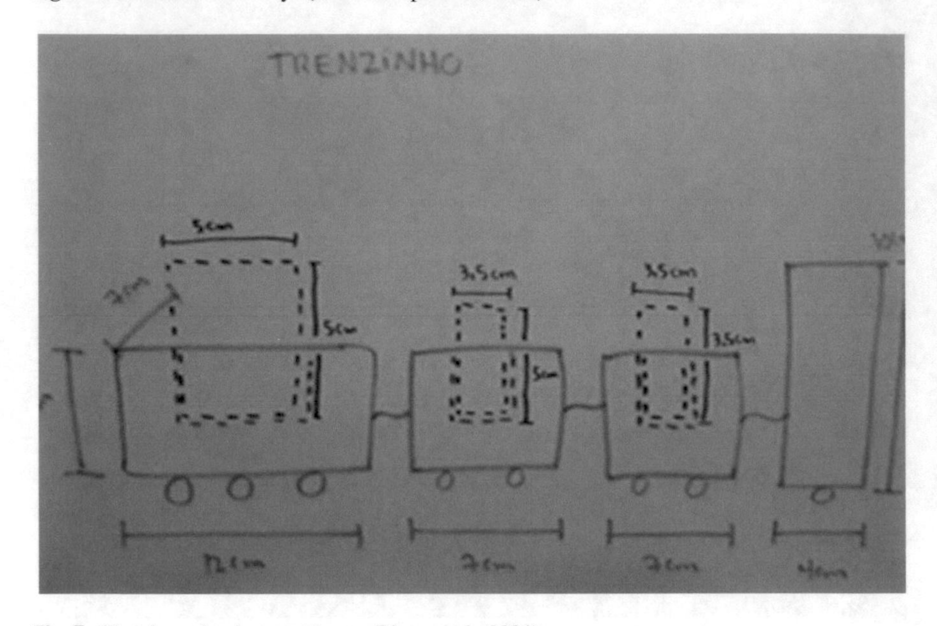

Fig. 7 Sketches of train toys (*Source* Bispo et al., 2021)

- Diameters above 15 cm: they have the potential to develop larger objects, such as bowling pins, trucks, trains, cars, robots, molds for houses and villages, planes, and tools (hammer, spanner, pliers, and saw).

The last phase of the toy manufacturing process is the construction of prototypes based on the sketches previously made. Some tools and utensils are essential at this stage, such as saws, hammer, screwdriver, and pocket knife (Bispo, 2017), which are responsible for the initial cuts and final finishes. Some examples of tools cited by Bispo (2017) can be seen below:

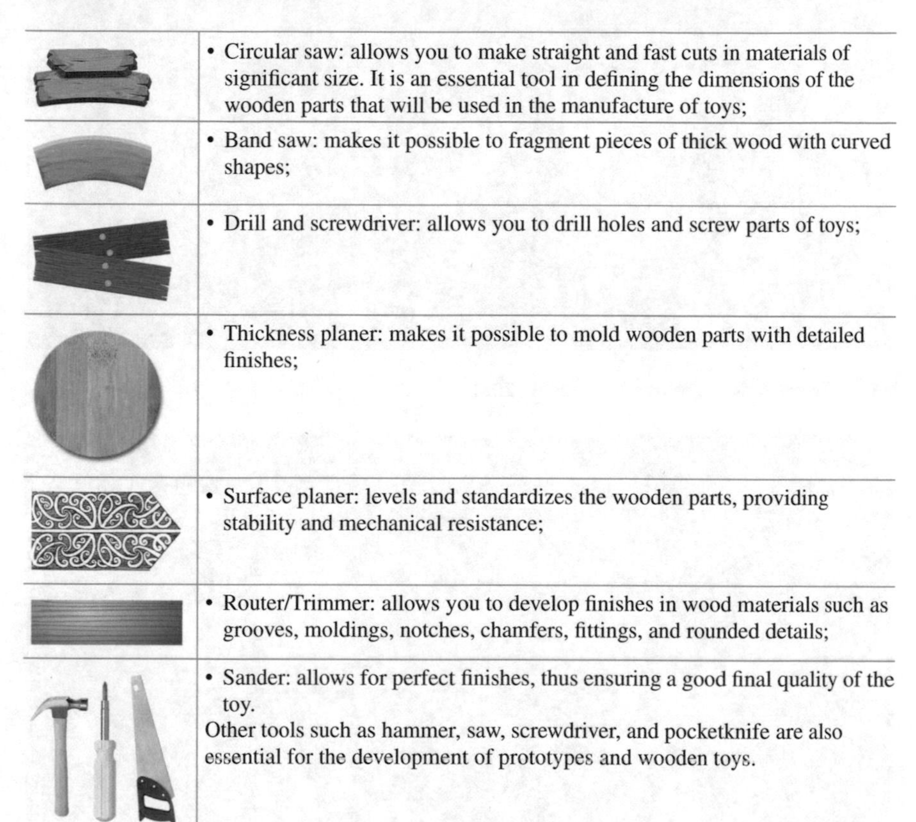

	• Circular saw: allows you to make straight and fast cuts in materials of significant size. It is an essential tool in defining the dimensions of the wooden parts that will be used in the manufacture of toys;
	• Band saw: makes it possible to fragment pieces of thick wood with curved shapes;
	• Drill and screwdriver: allows you to drill holes and screw parts of toys;
	• Thickness planer: makes it possible to mold wooden parts with detailed finishes;
	• Surface planer: levels and standardizes the wooden parts, providing stability and mechanical resistance;
	• Router/Trimmer: allows you to develop finishes in wood materials such as grooves, moldings, notches, chamfers, fittings, and rounded details;
	• Sander: allows for perfect finishes, thus ensuring a good final quality of the toy. Other tools such as hammer, saw, screwdriver, and pocketknife are also essential for the development of prototypes and wooden toys.

Toys that undergo machining process, such as those that simulate the reality of everyday life, are highlighted among children. Dolls, carts, and trains are some of the examples of children's products that fit into this niche and that can be manufactured using waste lignocellulosic materials from urban trees (Figs. 8, 9, 10 and 11). For the manufacture of more refined wood products, it is common to use glues, paints, and varnishes during the process of creating these objects. Bispo et al. (2021) emphasize that these inputs should not have any toxic chemical components in their composition. This is because the target audience is formed mainly by children. Thus, it is necessary to avoid products made up of solvents, resins, and/or toxic pigments that can trigger serious health problems.

Fig. 8 Wooden dolls (*Source* Bispo et al., 2021)

Fig. 9 Wooden colored carts (*Source* Bispo et al., 2021)

Fig. 10 Wooden trains with bark, color, and without bark (*Source* Bispo et al., 2021)

Fig. 11 Wooden carts (*Source* Bispo et al., 2021)

Fig. 12 Rustic bowling pieces (*Source* Bispo et al., 2021)

Because urban forestry wastes are mostly covered with bark, toys can also be created without the need to remove this tissue. As suggested, when they have larger dimensions, either in length and/or in diameter, bowling pieces are examples of rustic objects that can be manufactured (Fig. 12). In general, the barks can be used in toy designs with more vintage characteristics such as decorative elements of pieces such as house roofs, train sets, puzzles, and the external painting of carts (Fig. 13).

Not removing the bark prevents the generation of more waste, so that the collected materials are used completely, in addition to reducing the use of products such as paints and varnishes. However, because the bark is a region rich in extractives (Cristo et al., 2016; Rosell, 2019), previous analyses must be carried out in order to identify possible chemical compounds capable of causing an allergic reaction in the product handlers, especially kids.

4 Conclusions

The use of urban wood waste in the manufacture of toys has three main beneficiaries: the environment, the healthy development of children, and public management of urban solid waste. This application allows us to take the paths to be followed in favor of the sustainable development goals, ensuring the strengthening of the use of waste toward sustainable development from the reuse of renewable material of wide availability and low cost.

Fig. 13 Rustic carts (*Source* Bispo et al., 2021)

In this context, this raw material works as the central interface of the entire process of developing children's toys for different age groups, presenting as its main challenge the heterogeneity of its technological properties. However, this obstacle is easily overcome when the material is correctly characterized. Its manufacturing process is governed by a multidisciplinary team responsible for collecting the material and working on it until the final stages of finishing the toys.

Throughout the toy production process, design is an essential component, acting as an indicator of the user-product relationship. Therefore, there is the production of sketches, understood as a complex and dynamic process that is decisive in the prototype production phase; it is the phase of the process responsible for incorporating important characteristics to the product, such as attractiveness, child-toy interaction, ensuring specifics of the age group for which it is intended, in addition to aspects regarding user safety. Thus, the design allows the construction of an object that will help the child's cognitive, emotional, behavioral, and social development.

References

Arnstadt, T., Hoppe, B., Kahl, T., Kellner, H., Krüger, D., Bauhus, J., & Hofrichter, M. (2016). Dynamics of fungal community composition, decomposition and resulting deadwood properties in logs of Fagus sylvatica, Picea abies and Pinus sylvestris. *Forest Ecology and Management, 382,* 129–142. https://doi.org/10.1016/j.foreco.2016.10.004

Babich, M. A., Hatlelid, K. M., & Wanna-Nakamura, S. C. (2014). Toy safety and hazards. In *Encyclopedia of toxicology* (3rd ed., Vol. 4). Elsevier. https://doi.org/10.1016/B978-0-12-386454-3.00455-3

Bispo, L. F. P. (2017). *Aproveitamento de Resíduos da Arborização Urbana para a Fabricação de Brinquedos.* Universidade de São Paulo Escola Superior de Agricultura "Luiz de Queiroz."

Bispo, L. F. P., Nolasco, A. M., Souza, E. C., de Klingenberg, D., & Dias Júnior, A. F. (2021). Valorizing urban forestry waste through the manufacture of toys. *Waste Management, 126,* 351–359. https://doi.org/10.1016/j.wasman.2021.03.028

Bonsiepe, G. (1997). *Design: Do material ao digital.*

Cristo, J. S., Matias, E. F. F., Figueredo, F. G., Santos, J. F. S., Pereira, N. L. F., Junior, J. G. A. S., Aquino, P. E. A., Nogueira, M. N. F., Ribeiro-Filho, J., de A.B. Cunha, F., Costa, M. S., Campina, F. F., Tintino, S. R., Salgueiro, C. C. M., & Coutinho, H. D. M. (2016). HPLC profile and antibiotic-modifying activity of Azadirachta indica A. Juss (Meliaceae). *Industrial Crops and Products, 94,* 903–908. https://doi.org/10.1016/j.indcrop.2016.10.001

Cywa, K., & Wacnik, A. (2020). First representative xylological data on the exploitation of wood by early medieval woodcrafters in the Polesia region, southwestern Belarus. *Journal of Archaeological Science: Reports, 30,* 102252. https://doi.org/10.1016/j.jasrep.2020.102252

De Meira, A. M. (2010). Gestão de resíduos da arborização urbana. In *Tese (Doutorado em Ciência Florestal) - Universidade de São Paulo Escola Superior de Agricultura "Luiz de Queiroz."* Universidade de São Paulo, Escola Superior de Agricultura "Luiz de Queiroz."

de Meira, A. M., Nolasco, A. M., Klingenberg, D., de Souza, E. C., & Dias Júnior, A. F. (2021). Insights into the reuse of urban forestry wood waste for charcoal production. *Clean Technologies and Environmental Policy.* https://doi.org/10.1007/s10098-021-02181-1

Ebner, M., & Petutschnigg, A. J. (2007). Potentials of thermally modified beech (Fagus sylvatica) wood for application in toy construction and design. *Materials & Design, 28,* 1753–1759. https://doi.org/10.1016/j.matdes.2006.05.015

Faller, da R. (2009). *Engenharia e Design: contribuição ao estudo da seleção de materiais no projeto de produto com foco nas características intangíveis.* Universidade Federal do Rio Grande do Sul.

Ferrara, K., Hirsh-Pasek, K., Newcombe, N. S., Golinkoff, R. M., & Lam, W. S. (2011). Block talk: Spatial language during block play. *Mind, Brain, and Education, 5*(3), 143–151. https://doi.org/10.1111/j.1751-228X.2011.01122.x

Finžgar, D., Rupel, M., Humar, M., & Kraigher, H. (2014). Determining density and moisture content of the European beech (Fagus sylvatica L.) coarse woody debris from the secondary virgin forest Rajhenavski Rog. *Acta Silvae et Ligni, 103,* 35–46. https://doi.org/10.20315/ASetL.103.3

Goldstein, J. (2012). *Play in children's development, health and well-being* (Issue February). Toy Industries of Europe.

Guney, M., Kismelyeva, S., Akimzhanova, Z., & Beisova, K. (2020). Potentially toxic elements in toys and children's jewelry: A critical review of recent advances in legislation and in scientific research. *Environmental Pollution, 264,* 114627. https://doi.org/10.1016/j.envpol.2020.114627

Healey, A., & Mendelsohn, A. (2019). Selecting appropriate toys for young children in the digital era. *Pediatrics, 143*(1). https://doi.org/10.1542/peds.2018-3348

Kishimoto, T. M. (1995). O jogo e a educação infantil. *Revista Pro-Posições, 6*(2), 45–63.

Kishimoto, T. M. (2014). Jogos, brinquedos e brincadeiras do Brasil. *Espacios En Blanco, 24,* 81–106.

Klin, A., Jones, W., Schultz, R., & Volkmar, F. (2003). The enactive mind, or from actions to cognition: Lessons from autism. *Philosophical Transactions of the Royal Society B: Biological Sciences, 358*(1430), 345–360. https://doi.org/10.1098/rstb.2002.1202

Klingenberg, D., Nolasco, A. M., Dias Júnior, A. F., & de Souza, E. C. (2021). *Propriedades tecnológicas de sete espécies provenientes da arborização urbana* (pp. 305–316). https://doi.org/10.37885/201101966

Lee-Cultura, S., Sharma, K., & Giannakos, M. (2021). Children's play and problem-solving in motion-based learning technologies using a multi-modal mixed methods approach. *International Journal of Child-Computer Interaction.* https://doi.org/10.1016/j.ijcci.2021.100355

Levin, D. E., & Rosenquest, B. (2001). The increasing role of electronic toys in the lives of infants and toddlers: Should we be concerned? *Contemporary Issues in Early Childhood, 2*(2), 242–247. https://doi.org/10.2304/ciec.2001.2.2.9

Liu, R., Stamper, J., Davenport, J., Crossley, S., McNamara, D., Nzinga, K., & Sherin, B. (2019). Learning linkages: Integrating data streams of multiple modalities and timescales. *Journal of Computer Assisted Learning, 35,* 99–109. https://doi.org/10.1111/jcal.12315

Manzoor, J., Sharma, M., Sofi, I. R., & Dar, A. A. (2020). *Plastic waste environmental and human health impacts* (pp. 29–37). https://doi.org/10.4018/978-1-5225-9452-9.ch002

Nasser, R. A., Salem, M. Z. M., Al-Mefarrej, H. A., & Aref, I. M. (2016). Use of tree pruning wastes for manufacturing of wood reinforced cement composites. *Cement and Concrete Composites, 72,* 246–256. https://doi.org/10.1016/j.cemconcomp.2016.06.008

Nowak, D. J., Greenfield, E. J., & Ash, R. M. (2019). Annual biomass loss and potential value of urban tree waste in the United States. *Urban Forestry & Urban Greening, 46,* 126469. https://doi.org/10.1016/j.ufug.2019.126469

Ofer, N., Hitron, T., Erel, H., Zuckerman, O., & David, I. (2018). A little bit of coding goes a long way: Effects of coding on outdoor play. *Proceedings of the 17th ACM Conference on Interaction Design and Children*, 599–604. https://doi.org/10.1145/3202185.3210782

Okunola, A. A., Kehinde, I. O., Oluwaseun, A., & Olufiropo, E. A. (2019). Public and environmental health effects of plastic wastes disposal: A review. *Journal of Toxicology and Risk Assessment, 5*(2). https://doi.org/10.23937/2572-4061.1510021

Palharini, K. M., Guimarães Junior, J. B., Faria, D. L., Mendes, R. F., Protásio, T. D. P., & Mendes, L. M. (2018). Potential usage of the urban pruning residue for production of wood based panels. *Nativa, 6*(3), 321. https://doi.org/10.31413/nativa.v6i3.5418

Prata, J. C., Silva, A. L. P., da Costa, J. P., Mouneyrac, C., Walker, T. R., Duarte, A. C., & Rocha-Santos, T. (2019). Solutions and integrated strategies for the control and mitigation of plastic and microplastic pollution. *International Journal of Environmental Research and Public Health, 16*(13), 2411. https://doi.org/10.3390/ijerph16132411

Regal, P. H. A. (2003). A prática gráfica do croqui e a criatividade. *Revista Educação Gráfica, 7,* 19–32.

Rosell, J. A. (2019). Bark in woody plants: Understanding the diversity of a multifunctional structure. *Integrative and Comparative Biology, 59*(3), 535–547. https://doi.org/10.1093/icb/icz057

Rubens, C., Braley, S., Torpegaard, J., Lind, N., Vertegaal, R., & Merritt, T. (2020). Observations of children constructing and playing with programmable matter. *Proceedings of the Fourteenth International Conference on Tangible, Embedded, and Embodied Interaction*, 193–205.

Seghiri, M., Boutoutaou, D., Kriker, A., & Hachani, M. I. (2017). The possibility of making a composite material from waste plastic. *Energy Procedia, 119,* 163–169. https://doi.org/10.1016/j.egypro.2017.07.065

Sridhar, P. K., Nanayakkara, S., & Huber, J. (2017). Towards understanding of play with augmented toys. *ACM International Conference Proceeding Series.* https://doi.org/10.1145/3041164.3041191

TIE—Toy Industries of Europe. (2021). *Toys: The tools of play—Toy industries of Europe.*

Tiseo, I. (2021). *Global plastic production 1950–2020 | Statista.* Statista.

United Nations. (2021). *Sustainable development goals.* https://sdgs.un.org/goals

Viegas, V. A., de Pereira, S. J., Guimarães, K. D. L. M., Rocha, L. T. C., & Valporto, M. S. (2014). Propriedades das madeiras e suas relações com os requisitos de projetos: indicações de uso em brinquedos de madeira. *Blucher Design Proceedings, 1,* 1–12.

Vygotski, L. S. (1991). *A formação social da mente.* Martins Fontes.

Waksman, R. D., & Harada, M. de J. C. S. (2005). Escolha de brinquedos seguros e o desenvolvimento infantil. *Revista Paulista De Pediatria, 23*(1), 41–48.

Weisberg, D. S., Hirsh-Pasek, K., & Golinkoff, R. M. (2013). Guided play: Where curricular goals meet a playful pedagogy. *Mind, Brain, and Education, 7*(2), 104–112. https://doi.org/10.1111/mbe.12015

Yelland, N. (1999). Technology as play. *Early Childhood Education Journal, 26*(4), 217–220. https://doi.org/10.1023/A:1022907505087

Yu, J., Zheng, C., Tamashiro, M. A., Gonzalez-Millan, C., & Roque, R. (2020). Code attach: Engaging young children in computational thinking through physical play activities. *Proceedings of the Fourteenth International Conference on Tangible, Embedded, and Embodied Interaction,* 453–459. https://doi.org/10.1145/3374920.3374972

Zamora, A. M., Caterbow, A., Nobre, C. R., Duran, C., Muffett, C., Flood, C., Rehmer, C., Chemnitz, C., Lauwigi, C., Arkin, C., da Costa, C., Teles, D. B., Amorim, D., Azoulay, D., Knoblauch, D., Seeger, D., Moun, D., da Silveira, I., Patton, J., … Feit, S. (2020). *Atlas do plástico - Fatos e números sobre o mundo dos polímeros sintéticos* (Fundação Heinrich Böll [ed.]; 1st ed.).

Zosh, J. M., Verdine, B. N., Filipowicz, A., Golinkoff, R. M., Hirsh-Pasek, K., & Newcombe, N. S. (2015). Talking shape: Parental language with electronic versus traditional shape sorters. *Mind, Brain, and Education, 9*(3), 136–144. https://doi.org/10.1111/mbe.12082

On Longevity and Lost Toys: Sustainable Approaches to Toy Design and Contemporary Play

Katriina Heljakka

Abstract This chapter summarizes and synthesizes key developments within the industries of play and contemporary toy design that address sustainability in toys from two perspectives: first, from the design and production perspective, and second, from the use or play perspective. By drawing on insights from earlier work in the fields of toy and play research, a summary of key points regarding sustainable approaches to toy design and contemporary play is suggested. Good toy design aims at producing play value derived from raw material, aesthetics, mechanics, relations to storytelling and so on, but as a vague and subjective measure, play value can present different things for designers and players of different ages, depending on their ideas on the life cycles of toys and their qualifications for sustainability. Instead, relevant and more tangible aspects of play value are the physical durability of toys, which may refer to the ecological dimensions of the raw material used for toys (such as textile, wood or plastic), their functionality (afforded, versatile and enduring functions for toys such as mechanics) and the thematic durability of toys (such as intriguing backstories that invite the player to long-term play). When the toys' play value is considered thoroughly from these perspectives, the affective responses to the plaything are strengthened. In the chapter, the concepts of longevity and becoming lost are discussed as issues worth of deliberation not only within the toy industry, but in terms of player actions with toys.

Keywords Play value · Toys · Toy design · Toy play · Sustainability

1 Introduction: Where Are the Toys Going?

"Does this spark joy?" If it does, keep it, says the philosophy of Marie Kondo. On the one hand, toys are prime examples of items that are designed to 'spark joy' but often end up as waste when a child's play interests change (Robertson-Fall, 2021). On the

K. Heljakka (✉)
Degree Programme of Cultural Production and Landscape Studies (Digital Culture), University of Turku, Pori, Finland
e-mail: katriina.heljakka@utu.fi

S. S. Muthu (eds.), *Toys and Sustainability*, Environmental Footprints and Eco-design of Products and Processes,
https://doi.org/10.1007/978-981-16-9673-2_2

other hand, toys are the ultimate springboards of enjoyment and the imagination for players of many ages. "A toy is the embodiment of plasticity. The plastic toy can be throwaway or collectible; it can adapt to any context, and its low cost means that it can be produced in hundreds, or thousands, or tens of thousands, a multicolored, many-headed army spreading out to colonize the world with plastic culture" (Phoenix, 2006, p. 7).

Toy experiences are physical, fictional, functional and affective (Heljakka in Paavilainen & Heljakka, 2018). As either mass-produced or handmade objects toys fulfil their utilitarian role in play and may possess deep spiritual significance as objects that allow dialogues with the self. They may be used for decorative purposes and are thus regarded to have highly aesthetic value. Again, they can have deep significance for their owners as companions, muses and play partners. Emotional attachment is important when considering the wow-factor of the toy. Toys indeed, come to have emotional value for both the purchaser and the player. For example, the stories related to artefacts are important to developing a feeling of attachment (Fukuda, 2010, p. 3). But where do 'toy stories' begin and end, and where will the toys go, when they have fulfilled their cause as playthings?

The issues of longevity and becoming lost are relevant when considering toys and sustainability.

All products have a limited lifetime, toys more than most. Kids lose interest in particular toys and quickly move into new ones (Kline, 1993, p. 224). Rinker sees 'Trenders' as the individuals who own the "hottest" toys. He elaborates: They value toys not for their "playability", but for the prestige of ownership. The minute the craze is over, they are into something else. If you are a collector, this is a personality worth identifying. Since a trender's attention span is short, his toys usually don't show heavy use; because he has little attachment to them, he abandons them when leaving home. Eventually, parents simply want to get rid of them (Rinker, 1991, p. 6).

How do commercial toys relate to consumption? Many parents are interested in the utilitarian value of toys, namely their play value. Toys come with play value that stems of the playthings being educational, safe, durable, age-appropriate, ergonomic, simple, aesthetic, pleasurable, versatile, fun/amusing and sustainable (Heljakka, 2013). The complexity, variability, newness and novelty expected of toys are a product of capitalism (Calkin, n.d.).

Fousteri and Liamadis (2021, p. 590) claim that "open-ended toys are related to less consumption, as they can be replayed and re-used creating meaningful experiences, repeatedly", and that they "welcome the unknown, the unexpected and the unordinary". However, even sturdy, valued toys with non-precisive affordances for open-ended play may become broken, beyond repair and cannot anymore be used as manipulated play objects. Instead, they may continue their lives as decorative items, which endure, or are replaced with new toys and, consequently, become 'lost'.

Let us begin by exploring how toys can be seen as artefacts that on the one hand, arouse both a feeling of longing and on the other hand, represent the quality of longevity: The question of time and endurance in reference to toy culture is a complex one. Toys' time-related dimension refers to the many levels in the relations between the player and the toy. First, toys can be seen as time machines, time capsules and transformative objects in relation to how we view and experience time (Heljakka, 2013). Second, toys can be investigated as objects that offer long-term use. The toy's journey starts from design work, which is next taken into production. From the market place, the toys travel to the player realms, to those who use the toys for various play purposes and patterns for different lengths of relations and uses.

"How we dispose of the things we once loved is of utmost importance" (Robertson-Fall, 2021). At the end of the toy's life cycle lies the question of disposal, the toy's ultimate destiny; will it remain within the circular economy of playthings or dissolve into recycled material for some other purpose—find its way to a new player or, at worst, become waste dumped in landfills or waters?

A starting point for considerations on toys longevity or them becoming 'lost' is the area of toy design. Individual toy designers as well as the toy industry constantly feed the players of the world with new toys and ideas for play. Toy design entails the work of both amateur and professional toy designers, who design, model and create toys as part of individual craft businesses or as part of toy companies involved in the global ecosystem and industries of play (Heljakka, 2016b). In general, toy designers aim for novelty and uniqueness in their design work, which means that the realm of toys and their persuasive strategies to invite engagement and interaction with them are constantly renewed. In this way, the industries of play (Heljakka, 2013) largely operate in similar terms as the fashion industry, where designers and companies are insatiable for trends in emerging raw materials, production techniques, styles and by keeping an eye on the innovation in other areas of transmedia culture, such as art, television, movies, graphic design, comics and music. Currently and even increasingly so, when seeking inspiration for new playthings, the emphasis is put on player-generated ideas and content spread, shared and trending on social media.

Mass-produced toys come in terms of design in most cases form the U.S., Europe and Japan. In terms of production, the majority today comes from Asia, in most cases China. The largest volume comes from global giants such as Hasbro and Mattel. In reference to time, mass-produced toys are either bought as pristine, new products in hypermarkets or small-scale points of purchase, or as second-hand items in flea markets, garage sales, toy conventions and online shops such as eBay.

What is interesting in the context of this chapter is how toys 'mature' and what happens to them towards the end of their life cycle. Popular narratives, like the toy-tropes employed in the *Toy Story* franchise movies, play on the ideas that good toys are recycled and given the possibility to re-enter the sphere of play through engagement by new players. Again, the concept of longevity is employed in the chapter at hand in parallel to examples of toys to reflect on the ways certain toys are designed to endure time: Through a sustainable design process, they aim at both durability and endurance—a status of classics, rather than curiosities, which may finally become cult objects (Geraghty, 2014). As such, toys with long-term allure and play value acquire a seemingly permanent place in the toy market, in the players' hands and minds, and by a presence on the shelves and social media sites curated and cultivated by the enthusiasts of contemporary toy cultures.

However, according to Robertson-Fall (2021), 80% of all toys end up in landfill, incinerators or the ocean. For example, in France, some 40 million toys end up as waste each year. The chapter discusses the handicaps and fate of these lost toys—and what determines their ephemerality as objects lacking use value and therefore, lasting interest. The chapter continues with the idea of sustainability being embedded in various recyclable raw materials, functionality through engaging interactive affordances and thematic durability through captivating backstories, which are reinforced by designers' and players' commitment for longevity in design and playful interaction, and the innovativeness in coming up with ideas to recycle, replay and reappropriate toys so that they no longer become lost in the constant stream of novel playthings. Tackling overconsumption and endless consumerism are key.

This chapter continues with an exploration of perspectives on play and toys today and moves on to highlight toys as a part of the industries of play. Next, the similarities and differences in the life cycles of children's and adults' toy relationships are considered, followed by an introduction to the theme of sustainability within the toy sector. Here, an emphasis is made on production materials and chemicals, whereafter it is illustrated, how sustainable approaches have been made in toy design. The final part of the chapter summarizes and synthesizes approaches to sustainable play made perceivable by the industries of play and related news articles. To conclude the chapter, the author addresses the facets of the emerging circular economy of toys and ends by presenting key ideas on how environmentalism as part of sustainable thinking can lead to longevity in play, and fewer lost toys.

2 Perspectives on Play and Toys Today

The concept of play encompasses all forms of play and play with or without the player's own body, various play equipment and environments—physical or digital. In play, we explore possible pathways into the futur (Henricks, 2008, p. 174). Nowadays, both children and adults play a lot: People of all ages play with toys and play games, but children's toy play is the most recognized and legitimized form of object play. Children also play more spontaneously than adults with, for example, improvised toys made of natural materials (e.g. stones and sticks), while it is easier for adults to throw themselves into play with more ready-made tools, either handmade or industrially made toys. To illustrate differences in children's and adults' capabilities of using their imagination with a plaything, I often use the example of a cardboard box. If you give your child an empty cardboard box, he or she will likely quickly develop play patterns around it, but if you do the same for an adult, he or she will very likely have to draw a face or other clues on the box, causing the everyday object to communicate (more) playful meanings and invitations. In industry-produced toys, these offers are made by designers as affordances for play that are easier for adults to understand.

The first mass-produced toys appeared during the early industrial period (Fousteri & Liamadis, 2021, p. 595). In Western culture, play became more prominent for children with the Victorian era of the 1800s, with the developments of industrialization and urbanization. In current times, the play cultures of children and adults are getting closer to each other. The intergenerational aspect of play is a trend that has intensified in the 2020s. To demonstrate, adults and children consume similar entertainment content and transmedia phenomena (e.g. Star Wars, Pokémon, My Little Pony, etc.), which is why their toy preferences and play habits may also be consistent. Nevertheless, toys are experiencing evolutions. For example, My Little Pony in the 1980s looked quite different than it does today, and this may have consequences for player preferences and toy relations.

The central role of visuality of play is evident in cultural environments where players have access to smart devices, mobile technology and social media platforms. Toy play is at the same time an imagination-driven activity that involves the manipulation and modification of physical or digital (even hybrid) materials to suit the play patterns. What connects adults and children are their relationships to 'toyfriends' of various kinds. Traditional character toys, such as dolls, action figures and soft toys, are still popular, but smart toy robots and connected Internet of Toys characters are becoming more common in areas of both leisure and education, with which players can consume educational entertainment (edutainment). These toys are updated as they network with new content and provide learning tasks and in this way, continuous opportunities for toy-based learning. To their content, technologically enhanced toys are becoming more game-like, whereas games are becoming more toy-like. This trend is not new, but more visible as the material and digital worlds converge in many ways. For example, in games, more toy-like characters are emerging, and it is possible to engage in functions familiar from games. All the same, toys are invitations to play.

3 Toys Within the Industries of Play

Objects have always been used in play. According to semiotician and game researcher Mattia Thibault (2017), it is unlikely that we would find a civilization that did not have play equipment. The concept of a toy has undergone notable developments: In the past, the toy was used in many ways as a metaphor for various gadgets, technical devices and even means of transport. According to recent research, it is more common nowadays to think of all kinds of smart and mobile devices, from mobile phones to tablets, as self-serving toys (Ihamäki & Heljakka, 2018).

Commercial and mass-produced toys are the most used objects in modern play, and they are most often produced within the toy industry. The toy industry is a rapidly renewable sector, looking both forwards and backwards and today for both young and older consumers. The value of the global toy market exceeded USD 90 billion in 2019 (Robertson-Fall, 2021). This sector, which has traditionally focused on physical and analogue play equipment, both entertaining and educational, is at the same time an evolving industry, increasingly linked to technological developments, especially in digital communication technologies. The toy industry is, indeed, a fashion industry that lives of innovation, creativity and new product design. In toy design, the aim is to constantly apply the latest technologies and material innovations and to utilize the phenomena created by digital culture and, for example, social media platforms as a bank of ideas and as conceptual resources for toy design.

Designing for children in the Western world mostly means toys designed for the mass market. For adults, there is a taste for a wider array of toys sold, for example, on the vintage toy market (such as Ebay) and the areas of new designer toys and collectables. Moreover, the play equipment produced by the toy industry has been accompanied by an area of experimental toy design, where distribution takes place through communities and websites focusing on small product launches instead of the mass market. The best known of these is probably the online community and marketplace Etsy.com, which markets itself with the phrase "If it's handmade, vintage, custom or unique, it's in Etsy". While the traditional toy market is regulated with strict legislation, handmade toys sold on the craft market are not controlled in similar ways.

In addition to expanding target groups with diversifying tastes, the toy market also shows a more open and experimental approach to the raw materials used in toys. The choice of raw material for a toy matters first for the toy designer and second for the consumer/player, who ultimately decides how the toy product will be perceived on the market and used in play. Today, as it turns out, novel designer toys seem to take a turn towards the materials with historical significance—textiles, wood—even paper. Plastic, especially in the form of vinyl, continues to thrive, but designers are constantly coming up with new ideas about the materiality of toys. The most innovative toys represent combinations of the traditional wood and more contemporary plastics, or wood combined with, for example, magnetic metal. Vinyl is one of the most important raw materials used in the production of designer toys and especially in those toys targeted and preferred by the adult audience. Traditional materials are creatively combined with each other: wood with metal or organic rubber and, on the other hand, cardboard with information technology. At the same time, toys are being developed of more ecologically oriented materials, such as various combinations of recycled plastics. Durability derived from raw material has consequences for the life cycles of both toys and player relations with them.

4 Toy Life Cycles in Children's and Adults' Play

According to Nachmanovitch, who has written about the aesthetics of play, the "mismatch" between people and objects is a direct result of the aesthetics of artefacts. Why is it hard for many to commit to the objects they acquire? In his work *Free Play. In Improvisation in Life an Art*, Nachmanovitch describes our relationship to objects being broken because of their ugliness, not, for example, because they utilize plastic or electronics. Whereas the aesthetic of toys is an important matter, their value stems from dimensions beyond their visuality and materiality. One aspect is the toys' relationship storytelling.

The toy designers and companies are responsible for the creating, spreading, sharing and continuing of 'toy stories' through the products making, selling and marketing activities, but it is the players who are interacting, evaluating and assessing the impact of the toys as sustainable objects. Ultimately, it is the players, who according to traditional thought decide, whether the toy stories will cease or live 'happily ever after', through recycling activities.

"The practice of design involves the consideration of economic, engineering, cultural, gender, disability and environmental concerns. Designers must work to integrate features that meet the needs of as many users as possible" (The Center for Universal Design, 1997).

Sustainability is a concept of importance for toy designers as well. "Making toys that stimulate imagination is pivotal to maintain desirability" (Robertson-Fall, 2021). Good toy design, again, may result in both play with the artefact and a sustainable toy (Rassi, 2012). The ultimate toy would, according to Sheenan and Andrews (2009, p. 94), 'suit girls and boys equally, appeal to adults as primary purchasers, be available in the shops or online at less than 30 pounds (some 35 euros for a high-value toy), provide some educational effect, create enormous 'play value' via, e.g. interactivity expand or advance as the child grows up, be safe to use, allow easy storage, be recyclable, enable children of different ages to play simultaneously, develop intergenerational loyalty, benefit from peripheral products and community, support the many kinds of play: boost social skills, imagination, coordination and physical development. Additionally, the toy should be robustly made and should be FUN!'. It is possible to envision how meeting these design-goals would result in a toy with a long life cycle of use. Next, I will elaborate on the longevity of toys through the concept of life cycles.

Margolin explains the concept of the product cycle as follows: "Every product goes through a process of development and use that begins with its conception, planning, and manufacturing, moves to its acquisition and use, and ends with its disassembly or disposal" (Margolin, 2002, p. 46). Relevant questions when studying objects, in Appadurai's thinking, are where it comes from, who has made it and what kind of 'career' the object has had so far. What is also important is to consider how the thing's use changes with its age and what happens to it when it reaches the end of its usefulness (Appadurai, 1986, p. 66). Play value is a key indicator for good toys, but often complex to define. Even the most well-designed toy can get lost in the multitude of objects owned, curated and cherished among today's players. When the toy is capable to promote attachment in the player, however, the possibility of a long-term relationship is realized. The playability of the toy is a matter of good design, but even more so if the toy offers possibilities for repeated and long-term play.

The life cycles of toys in children's and adults' play may vary: "Children want this year's toys and this year's innovations, for those are the items that are talked about" (del Vecchio, 2003, p. 132). Children's toys are usually acquired on the mass market, toy chains, department stores and hypermarkets, and increasingly, through online commerce. Playthings for the youngest are still marketed through printed media, such as toy catalogues and print advertisements, and children receive most toys as gifts from parents and other adults. The values and motivation for children's play are often influenced by the thought that play forms developmental enjoyment for the children, and consequently, many toys are labelled under the edutainment label, meaning that they afford possibilities for kids to learn while being entertained. As compared with adult toy relations, children's attachment to toys is often temporary, as old toys are quickly replaced with new ones, and toys easily become 'throw away' items, often disposed without much afterthought.

According to research in the field of adult toy cultures (Heljakka, 2013), adults are, to one part, interested in acquiring the same toys as children, such as Lego bricks, Barbie dolls and Star Wars toys to name a few, but at the same time also interested in toys that are designed, produced and marketed as niche products (such as designer toys) by independent toy companies, and sold in online outlets such as Etsy, specialty stores, curated lifestyle and museum shops. In adult toy cultures, the marketing messages are spread through fan communities both offline (meet ups, fairs and other events in physical venues) and online, for example, in Facebook groups. Adults are known to participate in toy collecting and creative cultivation, such as customization and displaying practices, as well as toy photography and videography. Most adults spend time with toys during their leisure, but the notion of playbor connected to toy cultures, meaning creative media production and sharing of toy narratives, such as long-term, serial doll-dramas is not unknown in adult cultures either (Heljakka, 2021).

del Vecchio writes that 'adults often find comfort in the same old thing, and it takes effort to get them to try something new, kids are just the opposite; they embrace newness and shy away from the same old thing' (del Vecchio, 2003). Adult toy players seem to have less toys than many children do, unless they are 'hard-core' collectors. In other words, adults are more considerate when choosing toys for themselves. At the same time, they seem to be more ecological, as their relationships with toys are often long term. Finally, many adults invest quite a significant number of resources and time to engage with their toys, but this is rather well planned as opposed to the spontaneous play of children. Keeping the toys in good shape, valuing them and planning the passing on of one's toys are prevalent considerations and activities in adult toy cultures, besides trading and selling activities. In sustainable terms, adult toy relationships, in this way, seem to involve more deliberation.

5 Sustainability

"The word sustainability comes from 'to sustain', which essentially means to provide support and prolong or preserve something". Further, the Cambridge Dictionary defines sustainability as "the quality of being able to continue over a period of time". "More companies and governments are paying attention to sustainability than ever before, making it easier to make more informed and ethical choices" (Future Learn, 2021). For the past ten years or so, the global industries of play meaning the producers of traditional toys and board games have engaged with the issues related to environmentalism, meaning a movement and ideology that aims to reduce the impact of human activities on the earth and its various inhabitants. As commonly known, environmental thinking and a focus on sustainability may offer possibilities for new markets to emerge. In this way, growth and responsibility go hand in hand in sustainable thinking.

Ecological sustainability has been addressed by toy experts for almost a decade now. For example, Reyne Rice, a toy industry trend analysts, saw eco-friendly toys as a 'sustainable topic' in 2013 and believes consumers to carve out budgets that promote eco-consciousness and concern for our natural resources (Play it!, 2013). What eco-friendliness may mean in toys from the industry viewpoint are operational safety standards, eco-friendly practices, innovative green products, new green principles, reducing the carbon footprint, products safer and less toxic, tracking materials used in their products reducing waste and energy used (Auerbach, 2009).

This 'environmentalist' approach was communicated at the Nuremberg International Toy Fair—the world's largest toy show, in 2020, as follows: "A new economy is emerging, offering plenty of growth opportunities: sustainability and green concerns, digital mutation, altruism and recognition of differences" (Trend presentation at Nuremberg International Toy Fair, 2020). In practice, this means, for example, creating new toy ideas with upcycling (Nuremberg Toy Fair Press release, October 2021). Sustainable approaches in toy design have been conceptualized as activities with an aim to serve the common good by having a positive effect on society: "Toys that take a stand towards the matters of environment, inclusion, sustainability, health, well-being, equity have undoubtedly social impact" (Fousteri & Liamadis, 2021, p. 591).

In this way, sustainability is becoming ever more important in the toy sector. Sustainability is about understanding and reducing the sources that go into the products we use, what happens to their discarded packaging, and where they end up in the end. Niebelschuetz (2021) notes that, while sustainable toys are playing a major role towards more eco-consciousness child education, they will also quickly impact the toy industry—affecting both manufacturers and consumers. Sustainable supply chains and eco-friendly packaging are already major consumer concerns, and they have become noted in the toy industry.

5.1 Production Materials and Chemicals

Sustainability is a result of "the quality of causing little or no damage to the environment and therefore able to continue for a long time"; "Sustainable practice benefits the environment by conserving and looking after Earth's resources, preventing global warming and extreme weather, and protecting lives". Sustainable thinking encompasses materials, wastage, human treatment and working conditions (Cambridge Dictionary). As toys most often represent physical and material entities, the 'nature' and life cycles of their production are crucial when considering their sustainability.

The toy industry is regulated by strict rules and directives especially concerned with matters regarding the safety in use of the toy. Safety is one of the most addressed topics when designing toys for children. Sarah Monks, who writes about the history of toy-making in Hong Kong, argues: Possibly the biggest single challenge for the industry in recent years has been the introduction in its major markets of even more stringent toy safety standards, backed by mandatory regulations concerning toy design, labelling and permissible levels of potentially harmful substances, such as phthalate (Monks, 2011, pp. 180–181). Besides phthalates, lead paint and magnetics have been, during the past years in the business, listed as potential safety hazards in the industry (Oppenheimer, 2009).

Nevertheless, it is plastics that seems to provoke most concerns in the toy market. Plasticine was invented in 1897 by William Harbutt. It was first made commercially in 1900 (Lambert, n.d.). In the post-war twentieth century, plastic became an epitome for toys and its shiny surface something that communicated a completely novel approach to producing things for play. The material allowed mass-production that had been impossible before, but not all responses were in favour of this newly discovered raw material. For example, to Roland Barthes, the repetitious plastic forms represent the triumph of an 'inhuman' industrial feel over the innocent, more organic playthings of children's traditional culture (Kline, 1993, p. 146).

The technical toys of the 1970s and 1980s were mainly made of plastic. To give examples, polystyrene, polyethylene and other synthetic materials were used as raw material not only for cars and trains but also for building block kits. The use of new raw materials and changed aesthetic and pedagogical perceptions affected the appearance of the toy. If in the past an attempt was made to accurately copy the objects of the adult world, the shapes of toys now became more imaginary, and their colour solutions took the place of emotional richness (Heinimaa, 2005).

Plastic remains the main sustainability concern of many toy companies (Sánchez, 2021), as it is the most used material for toys. 90% of all toys are currently made from plastic, which can be a highly durable material but is often fragile, and broken toys are the first to be thrown away. The sheer volume of plastic toys in circulation may seem overwhelming. To illustrate, the Lego bricks sold in one year would circle the world five times (Sawaya, n.d.). In a recent study conducted by Aurisano et al. (2021), it was found that children in Western countries have on average 18 kg of plastic toys.

The volumes of toys in domestic space present potential safety hazards for children: The UN Environment Programme (UNEP) has published a report that finds 25% of children's toys contain harmful chemicals (Technical University of Denmark, 2021). Plastic toys include potentially toxic materials like BPA, phthalates, lead, formaldehyde, flame retardant chemicals and PVC (Garcia-Toledo, n.d.).

However, the question of raw material in toys extends beyond questions of eco-friendliness of the material and chemicals used—sturdy materials such as wood and the plastic used for Lego bricks are, in fact, considered to have a sustainable, durable quality compared to more easily breakable plastics. In fact, it is not only the aspect of their main raw material that makes toys a matter of constant critiques, but also the play patterns, or ways of engagement with the toys that produce waste in themselves. For example, the phenomenal 'unboxing' toys of recent years produce a huge amount of useless plastic waste. Today's toy and play cultures are increasingly influenced by social media phenomena, such as unboxing videos. In fact, the play patterns associated with unboxing, opening packaging, wrapping and boxes of many sorts, have inspired the industry to produce items with the element of surprise featuring as the key affordance of the plaything, and in due course, designing new products for this category of mystery toys, producing a lot of waste material through excess packaging materials. For example, blind-box toys, such as the LOL toy range by MGA Entertainment, including small-scale dolls and figurines, have been highlighted to generate a significant amount of waste.

One of the goals of the World Economic Forum for 2021 is to reduce single-use plastic by 80% and global waste by 15% (WEF goal, 2021). For businesses making new toys, thinking about the materials that go into those toys is vital to eliminate safety hazards associated with, for example, lead and magnetics in toys, waste and pollution. Further, the production of plastics is heavily dependent on fossil resources such as oil, coal and natural gas. Various toy companies now present toys made from bio-based plastics that are environmentally friendly (Nuremberg Toy Fair Press release, October 2021). New material alternatives, such as maize, rubber, bamboo, wool and cork, are constantly being sought and experimented with by the industry. Use of rubber, however, evokes previously unarticulated concerns regarding the volumes of toys produced, even as parts of toys. Perhaps surprisingly, according to an article in *BusinessWeek* in 2006, Lego could be considered the world's No. 1 tire manufacturer; the factory produces about 306 million tiny rubber tires a year (Sawaya, n.d.).

6 Sustainability by Toy Design—An Overview

The life cycles of toys have to do with both their material endurance and conceptual durability. The first dimension is easy to understand, but what is meant by conceptual endurance? As Margolin notes, contemporary design work regards longevity as a desired value in products (2002, p. 49). How durable a toy is designed to be, for example, can extend the time it is played with, and durability has as much to do with how desirable a toy is over time as its physical attributes. The conceptual map presented by Heljakka (2013) on the elements of wow-ness in toys illustrates how visual value of toys, such as the aesthetics, sensory value, affective/empathy value, anthropomorphic value, narrative value all together with the sustainable value encouraging emotional attachment together, forms a spectrum of important dimensions promoting toys' longevity.

One of the most important goals for contemporary toy design should be creating sustainable value in toys. The Universal Design Principles definition design applied to toys strives to that the (1) toy is appealing, (2) how to play with the toy is clear, (3) toy is easy to use, (4) toy is adjustable, (5) toy promotes development, and (6) toy can be played in different ways. In general thinking, meeting these aims will produce a toy that will have conceptual endurance. One historical toy that is exemplary in fulfilling the aforementioned criteria is Lego: Fortune magazine recognized Lego's success in 2000, when it named Lego the "Toy of the Century".

The more known an object is, the more likely it is that it will lose play value, claims Rasmussen (1999, p. 51). Therefore, toy designers should make the toy as versatile as possible. Ingredients for a 'growing toy', suggests Wachtel, consist of the following: fun to use, interesting for the child, safe and durable, stimulates creativity and imagination, encourages inquisitiveness and resourcefulness, a tool for learning, challenging, yet not frustrating, invites repeated use and longevity, involves child interaction, addresses developing needs such as nostalgia—that "Magic ingredient", cost-effective (Wachtel, 2012). The popularity of Lego bricks, according to many, results from their versatility. Lego, in fact, is a new toy every day (Hunter & Waddell, 2008, pp. 3, 12).

According to a theory presented earlier (Heljakka, 2013, 2016a), toy relations evolve over the phases of the first-hand reaction of being wowed by the toy, transform into flow, once the toy is employed in play, and eventually, produce an atmosphere of glow around the toy, once it has received marks of play on it. In sustainable toy relationships, the wow element in the plaything dissolves into flow, a situation where the potential playfulness of the toy is successfully released by the player again resulting in a pleasant experience in which time and space lose their meaning to the person at play. Flow may be described as something that occurs while the toy has successfully managed to fulfil its core function—to tempt the player into manipulating it. What the players then may add as value to the toy in their play activities is according to the philosophy of the continuum glow. The last phase of the life cycle of a toy before it comes to its end through disposal, or alternatively, starts afresh again by 'wowing' a new player, is described by a kind of afterglow that makes the owner of the toy cherish the artefact by keeping it clean and well preserved, perhaps as a decorative item in a home. Thus, the process from wow to flow to glow and then afterglow, when successfully realized in a toy, describes an ideal situation that toy designers should strive for when endeavouring to create a sustainable toy experience. Furthermore, designers would benefit from considering how the 'secondary-wow' (or, double-wow) could emerge in engagement with a plaything through the discovery of hidden, unknown and therefore surprising element in the toy (Heljakka, 2016a).

7 Approaches to Sustainable Play

According to previous research in the field of toys, most toy ranges now have a two-to-three-year lifespan (Fleming, 1996, p. 116), but the relationships with toys may be much longer (Heljakka, 2013). Sustainable toys afford longer life cycles, and their sustainability may come to mean other things than material sturdiness. Daniel Miller notes that "material culture matters because objects create subjects much more than the other way around. It is the order of the relationship to objects and between objects that creates people through socialization" (Miller, 2008, p. 287).

I suggest that the wow may be determined at least in how the player experiences certain factors (one or more) integrated into the design and play value of the playthings, such as visuality, materiality, cuteness, backstory and a possibility to become emotionally attached with the toy character. Moreover, it is more probable to end with sustainable toy design, when the dimensions of the physical, fictional, functional and affective components have been considered in the design work. Also, what I propose is that these factors, when designed into the toy through affordances, may result in long lasting and therefore more sustainable toy experiences. Players actualize the wow and may contribute to how the toy attains that certain glow. The aura of 'glow' helps toys to maintain their importance and status in time. As not all people want to preserve their toys (or the toys of their own children, for that matter), they either end up in the trash bin or enter the market again as second-hand goods. If they have that certain glow about them, evidently, new players will find these toys, most often at flea markets. Hence, the toys will endure.

The following part of the chapter summarizes approaches to sustainable play formulizing a synthesis of key ideas presented. Leaning on the ideas highlighted among toy industry experts and earlier toy research, I suggest that:

- *Sustainable play can be achieved by universal toy design*: One approach to the sustainability of play is the employment of principles defined within universal design: the concept of Universal Design (UD) for play "where toys are designed from the beginning to offer the same level of appeal, access, support and engagement to all users". This allows more players to engage with the toys designed and potentially produces more play value in due course.
- *Sustainable play can be achieved by using eco-friendly and recyclable raw materials for toys*: For example, LOL Surprise toys are set to be made with biodegradable plastic and paper in 2020. New Lego bricks are currently made from virgin ABS plastic; however, the company has set itself the target of making all its pieces from renewable or recycled materials by 2030. 'The plastics of the future' are bio-based plastics. For example, California-based Green Toys uses post-consumer recycled plastic (milk bottles and yogurt containers) to make new toys. In 2018, Lego started making a range of elements from sugarcane-derived polyethylene, a soft, durable and flexible plastic. Sugarcane is fast-growing, doesn't compromise food security and is sourced using guidance from WWF (Robertson-Fall, 2021). One approach to sustainable toy design is avoiding to use of plastics altogether and using other materials like solid wood, natural rubber and organic, eco-friendly fabric.
- *Sustainable play can be achieved by promoting recycling of toys, reusing and sharing models:* "For the countless toys that already exist, reuse and sharing models are pivotal to prevent them from becoming waste. This is one of the motivations behind the Lego® Replay initiative, which encourages owners to donate their used bricks to children's charities. The initiative is currently being piloted in the US and is one of the business's first steps towards a circular economy". And, "In California, US, Toy-Cycle has established a recommerce platform and consignment system that enables outgrown toys to be shipped directly to the company to be

sorted and resold. Likewise, French association Rejoué has also been collecting, cleaning, repairing and reselling used toys since 2012, so far saving 300 tons of toys from landfill" (Robertson-Fall, 2021). Play industry initiatives for a more sustainable future of toys include Lego Replay, which encourages consumers to pass along the toys that are no longer used. According to the Lego Group, 97% of its bricks are already kept or shared by owners (ibid.). Another example is toy subscription models and services, such as Hong Kong-based Baton and UK-based Whirli (ibid.).

- *Sustainable play can be achieved by developing repairing solutions for broken toys.* 3D printing technology manufacturer Dagoma has established Toy Rescue, providing spare parts for toys that are broken. Dagoma has created a library of 3D-printable files of commonly lost or broken parts from the most popular toys of the last 40 years. It includes dolls' arms, dinosaur tails, car wheels and countless other parts that can be printed with a filament appropriate for the original toy. Files can also be requested for parts that aren't currently available in the library (Robertson-Fall, 2021). However, not all broken toys are suitable for repair, toys should be recycled when they become worn or when the clear plastic becomes cloudy (Garcia-Toledo, n.d.).
- *Sustainable play can be achieved by toys that highlight social responsibility*: Toys are almost always instructive in some capacity, even without a primarily pedagogical intent (Calkin, n.d.). Toys that encourage children to be more socially responsible are a trend that is visible in the toy market. Toys "can open the way to two different narrations, one engaging social impact and another aiming for social change" (Fousteri & Liamadis, 2021, p. 1). Toys can help children to understand complex situations by simplifying and clarifying them and can, for instance, be used in play to explain topics such as environment and climate change (Nuremberg Toy Fair Press release, October 2021). This approach is conceptualized in previous research as toy-based learning (Heljakka & Ihamäki, 2018). Educational toys that teach about climate change and plastic pollution can range from emotional connectors such as a kid's favourite plush sea turtle that is struggling with plastic straws in its coral reef (Niebelschuetz, 2021).
- *Sustainable play can be achieved by incorporating environmental values to playtime*: "How kids will meet the challenges of the future will be fundamentally different from how they meet the challenges of today, and toys can play a major role in helping to reinforce the lessons they've learned" (Niebelschuetz, 2021). "Sustainability should be taught to young children so that they embed healthy practices into their everyday lives". Clearly, teaching children "about composting, recycling, caring about animals, and shopping second hand" offers avenues to think and act in a sustainable way. "Children are never too young to learn the importance of sustainability or begin adopting eco-friendly habits, and experts say it all starts with playtime" (Genious of Play Team, 2021). "Advocating for open-ended and abstract toys, it is argued that letting children imagine and create our world and be the designers of their future will undoubtedly be for the common good" (Fousteri & Liamadis, 2021, p. 1).

8 Conclusions

The resources of Planet Earth are limited. Therefore, toys should not be grown out in a fast pace, as Sutton-Smith points out. "Instead, toys should grip the imagination and the imagination must grow them" (The Paradoxes of Toys, n.p.). This applies to toys used in play by players of all ages.

Ultimately, the goal is thus to develop this understanding so that more sustainable relationships between toys and players could be formed, and that toys would again be consciously designed so that they can either be recycled from player to player, or so that their raw material can be utilized more easily in the production of either new toys or other artefacts.

In our ever-expanding material world, one comes to wonder to what extent people are willing to become attached and to form sustainable relationships to artefacts. Nachmanovitz (1990, p. 150) brings up the question of people's 'nonrelation' with things. This means that products are to be considered bad in the sense that they do not encourage their owners to form long-term relations to the product. How to strengthen the sustainability, for example, through the play value in a toy, and in that way prolong play. "Toys should provide children with some immediate success. At the same time, they should allow the player to longer term engagement (perhaps weeks or months) to explore and understand the full complexity of the toy" (Sutton-Smith, The Psychology of Toys and Play, p. 3). Elkind (2007, p. 18) claims that if children have too many toys and have access to them all the time, they find it difficult to become attached to the toys and select the ones that stimulate the fantasy and imagination. In other words, even stand-alone toys should afford a long-term possibility of exploration in that way to cater for (continuous) curiosity. Challenges arise when toys become lost in the multitude of objects. Many adults and parents often ask: "How many toys does a child really need?" And I continue by asking: "How many toys does anyone need?".

In October 2021, two schoolchildren aged 10–11 years wrote in a Finnish newspaper opinion article about how climate change is a serious matter, and how we can stop it. They argued: "I can influence climate change myself by thinking: "Do I need new toys?"" (Merimets & Mäkelä, 2021). Indeed, today's children are not only more aware of the risks of climate change than their parents ever were in their youth, but also seem proactive in ways of negotiating, what the situation with the environment should mean for their own consumption, even in terms of toy shopping.

According to some views, a child does not need any commercial toys to support its development. Thinking about the values a family has, it could be considered what is an appropriate number of toys. Professor of early education, Kristiina Kumpulainen, observes that the concept of toy is wide. A child needs some toys, but play can be derived from everyday objects, such as cardboard boxes, sticks, rocks and kitchen bowls. Toy preferences of each time are dependent on the environment and culture. A child does not need specific toy types. Toys are surrounded by many questions related to value: The example of an adult is important as a child mirrors adult behaviour. Questions related to use of space, a peaceful atmosphere, possibilities for play and preferences for certain aesthetics are relevant when considering the number of toys in a home. Children should be involved in the conversations about toys and their uses (Kuluttavia kysymyksiä, 2021).

The guidelines given by Kumpulainen regarding toy volumes in a home are to pack one part of the toys away, to negotiate with the family about the right number of toys in the family and to converse about what is important: What kind of everyday life would the family like to spend, and what consumer choices can be done. For toys, this means asking the children about the kinds of toys they enjoy and how they can be positioned in the home. And are there toys in the children's room, which are no longer used—and what could be done to them? (Kuluttavia kysymyksiä, 2021).

Kumpulainen claims that questions regarding raising children connect with values and ethical choices, and not only mechanical deliberations on the volume of toys. The environmental burden of toys is a significant question, which could be used in raising future citizens. Questions that could be asked are what kind of ecological footprint each toy leaves, where toys come from, can they be fixed and where are the toys going when they are no longer needed. A discussion about toys' materiality and sustainability is a necessary topic in teaching ecological thinking (Kuluttavia kysymyksiä, 2021).

Toy ownership, toy production and play patterns are the cornerstones of the circular economy of toys, as the toymakers think about the future of the industries of play. The key goal of sustaining a circular economy is keeping the materials and products in use:

> The circular economy is a model of production and consumption, which involves sharing, leasing, reusing, repairing, refurbishing and recycling existing materials and products as long as possible. In this way, the life cycle of products is extended. In practice, it implies reducing waste to a minimum. When a product reaches the end of its life, its materials are kept within the economy wherever possible. These can be productively used again and again, thereby creating further value. (European Parliament, 2021)

When discussing toys and time, one may refer to the manifold dimensions of experiencing time in terms of the plaything and the playing activity. Topics of discussion may then relate to when and how we pass time with the toy, what is the toy's ability to last in time, and furthermore, how a toy affords itself as a material container for memories. Thus, the toy, understood in contemporary cultures, is an artefact that conceptually affords examination on many levels and on a temporal axis between the past and now. On the one hand, we may ask how a toy manages to awaken nostalgia for things as they were. On the other hand, perhaps more interestingly and seen from the design perspective, we might ask how to design sustainability in a toy so that it will have as long a life cycle as possible (Heljakka, 2013).

As questions about longevity and sustainability will be stressed in the material culture and industrial production of toys, one possibility lies in the chances of using yesterday's plastic pieces no longer attracting the player as a plaything, to produce something else of the toy, or perhaps, a completely new toy. Future toy designers will thus have many opportunities to come up with new ideas for 'old' toys in order to prolong their life cycles on the market. What has been a toy can become another type of a playful product. Alternatively, the toy can become raw material for something else. At the same time, mass-produced toys do not lose their aura because of reproduction, but maintain it exactly because of that. Therefore, 'toy stories' that come in volumes are still needed. What the designers should pay attention to focusing on their long-term play value, however, is to ensure that even mass-produced toys enable customization and personalization (and reparation!).

One of the developments I have also noted during the years of doing toy research is the wider toyification of culture, for example, in the fields of art and design. Artists are increasingly using toys in their making as both a raw material and a source of inspiration. In the design, this can be seen, e.g., as toy-like furniture and fashionable accessories and dresses with aesthetics clearly familiar from toys' shapes, colours and figures. These developments mean that more material resources are processed and used in shaping toy-like objects, artefacts and experiences for growing consumer groups. From another viewpoint, this promotes possibilities for actual toys to be recycled in the realms of art and design.

While toy designers of the modern age have needed to pay special attention to the aspects of safety, quality and durability, the ongoing ecological crises call for measures in terms of responsibility, renewable and recycled resources. At the time of writing the chapter (October 2021), Nuremberg International Toy Fair announced their 'mega trend' for year 2022, namely "Toys go Green", which seems to promote positive developments for the sustainability of toys. In consultation with the experts—the 13 members of the Trend committee—the team has identified four product categories relevant to the trend: "Made by Nature", "Inspired by Nature", "Recycle & Create" and "Discover Sustainability" (Nuremberg Toy Fair Press release, October 2021). And as for the toy players themselves, as well as the educationalists and parents of young children, their current and future toy literacy (Sutton-Smith, 1986) should also encompass 'environmentality'—striving to more 'zero-waste play' and longevity and, consequently, fewer lost toys.

Acknowledgements This research is conducted in affiliation with Pori Laboratory of Play.

References

Appadurai, A. (Ed.) (1986). *The social life of things. Commodities in cultural perspective.* Cambridge University Press.

Auerbach, S. (2009). *The art of the toy. A view and analysis for toy design.* Unpublished Research Paper (received from the author).

Aurisano, N., Huang, L., Milà i Canals, L., Jolliet, O., & Fantke, P. (2021). Chemicals of concern in plastic toys. *Environment International, 2021*, 146, 106194. https://doi.org/10.1016/j.envint.2020.106194

Calkin, T. (n.d.). *Whimmydiddles, whirligigs, and capital punishment: A history of toys and games, being a partial and idiosyncratic exploration of several centuries of developments, focusing largely on Europe and North America.* http://viralnet.net/essays/tylercalkin.html

Cambridge Dictionary. (2021). *Sustainability.* https://dictionary.cambridge.org/dictionary/english/sustainability

del Vecchio, G. (2003). *The blockbuster toy! How to invent the next BIG thing*. Pelican Publishing Company.

Elkind, J. (2007). *The power of play*. Da Capo Press.

European Parliament. (2021). *Circular economy: Definition, importance and benefits*. https://www.europarl.europa.eu/news/en/headlines/economy/20151201STO05603/circular-economy-definition-importance-and-benefits

Fleming, D. (1996). *Powerplay*. Manchester University Press.

Fousteri, A., & Liamadis, G. D. (2021). Toy stories for the common good. In M. Botta & S. Junginger (Eds.), *Swiss Design Network Symposium 2021 Conference Proceedings "Design as Common Good: Framing Design through Pluralism and Social Values"* (PP. 584–599). Online conference, 25–26 March.

Fukuda, S. (2010, November 29–December 1). Creative consumers and how we can meet their expectations. In *Proceedings of the first international conference on design creativity (ICDC2010)*. Kobe, Japan.

Future Learn. (2021). *50 of the biggest sustainability questions answered*. https://www.futurelearn.com/info/blog/50-sustainability-faqs-answered/

Garcia-Toledo, A. (n.d.). *Toxic materials you'll be shocked to find in toys*. The Tot. https://www.thetot.com/baby/toxic-materials-youll-be-shocked-to-find-in-toys/

Genious of Play Team. (2021). *4 tips to raise earth-conscious kids*. https://thegeniusofplay.org/genius/expert-advice/articles/4-tips-to-raise-earth-conscious-kids.aspx#.YXLTe55Bxdg

Geraghty, L. (2014). *Cult collectors. Nostalgia, fandom and collecting popular culture*. Routledge.

Heinimaa, M. (2005). *Sankarilliset lelut* [Heroic toys]. Atena kustannus Oy.

Heljakka, K. (2013). *Principles of play(fulness) in contemporary toy cultures. From wow to flow to glow*. Doctoral dissertation. Helsinki: Aalto University.

Heliakka, K. (2016a). Contemporary toys, adults, and creative material culture: From wow to flow to glow. In A. Malinovska & K. Lebek (Eds.), *Materiality and popular culture: The popular life of things* (pp. 249–261). Taylor & Francis.

Heljakka, K. (2016b). Strategies of social screen play(ers) across the ecosystem of play: Toys, games and hybrid social play in technologically mediated playscapes. *WiderScreen*, 1–2/2016. http://widerscreen.fi/numerot/2016-1-2/strategies-social-screenplayers-across-ecosystem-play-toys-games-hybrid-social-play-technologically-mediated-playscapes/

Heljakka, K. (2021). From playborers and kidults to toy players: Adults who play for pleasure, work, and leisure. In M. Alemany Oliver & R. W. Belk (Eds.), *Like a child would do. An interdisciplinary approach to childlikeness in past and current societies* (pp. 177–193). Universitas Press.

Heljakka, K., & Ihamäki, P. (2018). Preschoolers learning with the internet of toys: From toy-based edutainment to transmedia literacy. *Seminar.net, 14*(1), 85–102. https://journals.hioa.no/index.php/seminar/article/view/2835

Henricks, T. (2008, Fall). The nature of play. An overview. *American Journal of Play, 1*(2), 157–180.

Hunter, R. Jr., & Waddell, M. E. (2008). *Toybox leadership. Leadership lessons from the toys you loved as a child*. Thomas Nelson.

Ihamäki, P., & Heljakka, K. (2018). Smart toys for game-based and toy-based learning. A study of toy marketers', preschool teachers' and parents' perspectives on play. *The Eleventh International Conference on Advances in Human-Oriented and Personalized Mechanisms, Technologies and Services, CENTRIC*, 14–18.

Kline, S. (1993). *Out of the garden: Toys and children's culture in the age of TV marketing*. Verso.

Kuluttavia kysymyksiä. (2021). Paljonko leluja lapsi tarvitsee? [How many toys does a child need?]. *Lapsemme* 3/2021 *Kuluttavia kysymyksiä*. https://www.mll.fi/lapsemme-lehti/paljonko-leluja-lapsi-tarvitsee/

Lambert, T. (n.d.). *A brief history of toys*. https://www.arts.unsw.edu.au/sites/default/files/documents/GERRIC_The%20Games%20People%20Play%20Pre-Reading%202020.pdf

Margolin, V. (2002). *Politics of the artificial*. The University of Chicago Press.

Merimets, M., & Mäkelä, M. (2021, October 8). Ilmastonmuutos on vakava asia, me voimme pysäyttää sen [Climate change is a serious thing, we can stop it]. *Satakunnan Kansa Mielipide*, B2.
Miller, D. (2008). *The comfort of things*. Polity Press.
Monks, S. (2011). *Toy Town. How a Hong Kong industry played a global game as told to Sarah Monks*. Toy Manufacturers' Association of Hong Kong by PPP Company Ltd.
Nachmanovitz, S. (1990). *Free play: Improvisation in life and art*. Penguin/Tarcher.
Niebelschuetz, M. (2021). *The importance of sustainability in play*. https://thegeniusofplay.org/genius/expert-advice/articles/the-importance-of-sustainability-in-play.aspx#.YYlCT55Bxdg
Nuremberg International Toy Fair. (2020). Trend presentation at Spielwarenmesse 2020.
Oppenheimer, J. (2009). *Toy monster. The big, bad world of Mattel*. Wiley.
Paavilainen, J., Heljakka, K., Arjoranta, J., Kankainen, V., Lahdenperä, L., Koskinen, E., Kinnunen, J., Sihvonen, L., Nummenmaa, T., Mäyrä, F., & Koskimaa, R. (2018). *Hybrid social play final report*. Trim Research Reports (26).
Phoenix, W. (2006). *Plastic culture. How Japanese toys conquered the world*. Kodansha International.
Play it! (2013, February 2). A feel for trends. Daily News by Das Spielzeug, Spielwarenmesse International Toy Fair Nürnberg publication, 30.01.-04.02.2013.
Rasmussen, T. H. (1999). The virtual world of toys—Playing with toys in a Danish preschool. In Berg, L.-E., Nelson, A., & Svensson, K. (Eds.), *Toys in educational and sociocultural context. Toy research in the late twentieth century*. Part 2. Selection of papers presented at the International Toy Research Conference, Halmstad University, Sweden June, 1996, 47–57.
Rassi, J. (2012). "Hyvät, pahat ja hassut lelut" [The good, the bad and the funny toys]. Exhibition texts, Suomen Lelumuseo Hevosenkenkä, 21.3.2012–10.3.2013.
Rinker, H. L. (1991). *Collector's guide to toys, games and puzzles*. Wallace-Homestead Book Company.
Robertson-Fall, T. (2021, February 10). *Creating a circular economy for toys. Solutions to design out waste and pollution*. Ellen MacArthur Foundation. https://ellenmacarthurfoundation.org/articles/creating-a-circular-economy-for-toys
Sánchez, S. (2021). *TOY BY TOY. Understanding the challenges and opportunities that sustainability presents for the toy industry*. https://sonia-sanchez.com/toy-industry/
Sawaya, N. (publication year unknown). *Art of the brick*. Quirky LEGO facts. https://www.fi.edu/sites/default/files/PressKit_ArtOfTheBrick_Quirky_LEGO_Facts.pdf
Sheenan, M., & Andrews, P. (2009, February 14–27). Future toys. *Engineering & Technology, 2009*, 94–95.
Sutton-Smith, B. (unknown release date). The paradoxes of toys. Notes accessed at the Brian Sutton-Smith archives, The strong national museum of play.
Sutton-Smith, B. (unknown release date). The psychology of toys and play. Press release. Harshe-Rotman & Druck Inc.
Sutton-Smith, B. (1986). *Toys as culture*. Gardner Press.
The Center for Universal Design. (1997). *The principles of universal design*. Version 2.0. North Carolina State University.
Technical University of Denmark (2021, February 22). Potentially harmful chemicals found in plastic toys. *Science Daily*. www.sciencedaily.com/releases/2021/02/210222124552.htm
Thibault, M. (2017). *The meaning of play. A theory of playfulness, toys and games as cultural semiotic devices*. Unpublished doctoral dissertation, University of Turin.
Wachtel, P. (2012, February). Toys that grow with kids. Gifts and Decorative Accessories, No. 112.
Walker, S. (2006). *Sustainable by design. Explorations in theory and practice*. Earthscan.
WEF Goal. (2021). World economic forum. *Sustainable development impact summit*. https://www.weforum.org/events/sustainable-development-impact-summit-2021

Tips for Selecting Wood from Urban Afforestation for the Production of Toys: How the Sustainable Reuse of Waste Can Result in Economic, Environmental and Social Benefits

Elias Costa de Souza, Álison Moreira da Silva, Adriana Maria Nolasco, João Gilberto Meza Ucella-Filho, Regina Maria Gomes, Graziela Baptista Vidaurre, Rafael Rodolfo de Melo, Alexandre Santos Pimenta, José Otávio Brito, and Ananias Francisco Dias Júnior

Abstract Wood wastes from urban afforestation, which are commonly used in applications with lower added value, have interesting characteristics that can allow and facilitate their use in other sectors that use solid wood as a raw material for product development. Thus, the purpose of this chapter was to discuss the main aspects of wood quality that make it suitable for use in the production of small wooden objects, such as toys. For this, we discuss the mechanical, physical, and chemical properties and the ideal quality parameters that wood should have to validate its use for this purpose. Based on this, we suggest strategies for classifying and reusing wood waste from urban areas, so that these residues can be reused in the generation of higher value-added products. Thus, it is possible to combine the economic and social benefits with the environmental benefits of the correct disposal of this waste in different cities, which generates a significant amount of these materials. The main direct benefits arising from the use of urban wood waste for toy production can be divided into three main groups: environmental, through the reduction of polluting gas emissions, due to the burning and incorrect disposal of wood; economic, as the costs involved in the management of urban trees can be deducted from the value of the sale of wood for the production of toys, or even the direct sale of toys through industries managed

E. C. de Souza (✉) · Á. M. da Silva · A. M. Nolasco · R. M. Gomes · J. O. Brito
"Luiz de Queiroz" College of Agriculture, University of São Paulo, Piracicaba, SP, Brazil

J. G. M. Ucella-Filho · G. B. Vidaurre · A. F. Dias Júnior
Department of Forestry and Wood Sciences, Federal University of Espírito Santo—UFES, Jerônimo Monteiro, ES, Brazil

R. R. de Melo
Department of Agronomic and Forestry Sciences, Federal Rural University of the Semi-Arid Region, Mossoró, RN, Brazil

A. S. Pimenta
Academic Unit Specialized in Agricultural Sciences, Federal University of Rio Grande Do Norte, Macaíba, RN, Brazil

S. S. Muthu (eds.), *Toys and Sustainability*, Environmental Footprints and Eco-design of Products and Processes,
https://doi.org/10.1007/978-981-16-9673-2_3

by the municipalities; and social, with the direct generation of new jobs from the creation of this market opportunity, in addition to easier access to wooden toys, at a reduced cost, when compared to toys produced by large industries.

Keywords Urban wood waste · Wooden toys · Wood properties · Sustainable toy production · Sustainable development goals · Job creation

1 Introduction

Urban forests are present in almost every city around the world (Kampelmann, 2021). According to the region, different species, with specific characteristics, are planted in parks, squares and even on the sidewalks in front of the houses. Some cities invest significant amounts of money in urban afforestation, ensuring direct and indirect benefits to the population, such as maintaining the microclimate, decreasing temperature and sunlight, among other benefits related to well-being (Nowak & Dwyer, 2000; Nowak & Greenfield, 2018). However, for the good maintenance of these trees in urban spaces, some measures must be taken to ensure their integrity and health. One of the most common techniques is pruning, which consists of removing branches and can be performed to avoid damage to the physical structure of the places where the trees are planted and to ensure the safety of electrical networks present in urban spaces (Nowak et al., 2019; Ow et al., 2013). In some more complex cases, it is also necessary to fell and remove the trees (Nowak et al., 2019).

These urban tree management techniques generate a large amount of waste, both timber and non-timber, such as leaves, flowers, wood, among others (Kampelmann, 2021). Taking into account only the amount of dry wood from urban afforestation, countries like the United States generate about 33 ton every year (Nowak et al., 2019). These generated wastes are generally used in composting, disposed of in landfills, applied in energy production, or even burned in the open in developing countries (Meira et al., 2021; Joshi et al., 2015). In addition to being alternatives with low financial and social returns, some of these activities are harmful to human health and the environment, generating greenhouse gases, which end up contributing to the advancement of climate change (Fetene et al., 2018; Meira et al., 2021). However, some alternatives to these more traditional practices are already being studied, such as the use of urban wood waste in the production of small wooden objects, such as toys (Bispo et al., 2021).

Wooden toys are one of the most traditional forms of access to leisure and fun for children and have been known for hundreds of years in some cultures (Bispo et al., 2021; Cywa & Wacnik, 2020). In addition to leisure and fun, toys can play a pedagogical role, helping children's development, both in the physical-motor aspect and in the creative-intellectual aspect. Previous studies indicate that urban wood waste has some characteristics of interest for use in the production of toys (Bispo et al., 2021). It can contribute to sustainable social development, combining job creation with environmental sustainability, and can become a model to be followed in several cities around the world, thus helping to directly or indirectly achieve the goals 8 (decent work and economic growth), 11 (sustainable cities and communities), 12 (responsible consumption and production) and 13 (climate action) of the sustainable development goals (SDGs), adopted by the United Nations (United Nations, 2015).

However, in order to be used in toy production, wood must have some specific characteristics, which can be verified through the characterization of the species, which will allow the correct use of each species in the production of specific types of toys (Bispo et al., 2021). Some woods need to have higher density, for example, to be used in the production of toys that suffer greater impacts; in addition, the woods need to have a low content of total extractives, aiming at the safety of children who have direct contact with the toys (Bispo et al., 2021).

Although there are some indications about the main characteristics that wood must have to be used in the production of toys, the literature is still scarce, which makes it difficult to advance the development of projects in different cities around the world. Thus, the purpose of this chapter was to discuss the main aspects of wood quality that make it suitable for use in the production of small wooden objects, such as toys. Initially, we discussed the main properties of wood that can influence the production of toys. Next, we discuss strategies for classifying and reusing urban wood waste. Finally, we address the main economic, social and environmental benefits that can be achieved with the reuse of urban wood waste.

2 Properties Favorable to the Production of Toys

The properties of wood can directly influence the acceptance of a toy. Toys that emit a smell and have a bad taste, for example, can cause discomfort in children, logically, making the toy not the most desired. Also, aromatic compounds can be inappropriate for toys, as children can be intoxicated. Likewise, it is not desirable for a wooden toy to be fragile, making it useless in a short time, or retaining embossments that can hurt children. Next, the properties that are most interesting from the point of view of the quality of wooden toys are explained.

2.1 Anatomical Properties

The anatomical properties of wood constitute an important point to be considered when using species from urban tree pruning. This type of material has the most varied characteristics and, when it comes to toys, they can be decisive for the acceptance, or not, by children. Organoleptic properties, for example, are associated with characteristics perceptible through the human senses, without the use of instruments. Among the most observed properties in wood, applied in a practical way to the development of toys, are: color, smell, taste, texture, shine and design.

The color, resulting from the impregnation of substances from the cells and tissues of wood, as well as the design, texture and natural durability, can increase the options for using wood, making it possible to add natural beauty to toys. However, the colors of wood can change, due to variations in moisture content and temperature, attacks by woody organisms and photochemical reactions that occur in its structure (Pournou, 2020; Zabel & Morrell, 2020). Thus, heartwood color must always be observed on a tangential longitudinal surface, recently exposed, so that it is not altered by external factors.

Shine and design are also aesthetic characteristics of wood and give beauty to the toy, depending on the species to be used. The origin of the design that composes the wood comes from the interaction of the characteristics of the sapwood, heartwood, growth rings, grain, rays, fibers and, especially, the cutting plane to which the wood was subjected. Smell and taste are mainly due to volatile and/or soluble substances (Morata et al., 2019), which are mainly concentrated in the core. Its presence can enhance or limit the use of wood, especially when associated with toys intended for children. Naturally, children have the habit of putting objects in their mouths, as it is their starting point for getting to know the world at birth. A toy, no matter how beautiful, with striking and vivid colors, if it does not have a pleasant smell and taste, will be forgotten by the child.

Texture, which varies from fine to coarse, despite being noticeable to the naked eye, is defined through microscopy. This property is associated with the variation in the percentage, distribution and dimensions of adaptable wood elements (de Morais & Pereira, 2015). To suit children's preferences, finely textured woods are most desired. The hands of children, especially those under three years of age, are fragile and delicate, which makes the roughness of thick-textured woods liable to cause injuries, just by handling the toy, raising an alert about this property.

2.2 Chemical Properties

The cell wall is composed of macromolecular constituents: hemicelluloses, cellulose and lignin. These materials represent 95% of the dry weight of wood and define the shape, strength and structure of the material. In smaller proportions, there are strange compounds that perform accessory functions, representing around 5% of the total weight of the wood (Migneault et al., 2015; Murphy et al., 2021).

Cellulose is the most abundant component, representing 45% of the dry weight of wood. Hemicelluloses are amorphous elements that represent around 25–30% of the weight of the wood substance and are formed by polymer bonds between 10 monosaccharides (sugars) of various types, which is why it is a polyose (Lousada, 2020; Zhang et al., 2021). The set of hemicellulose and cellulose polymers makes up the total content of polysaccharides in the wood called holocellulose. Lignin, in turn, differs significantly in structure from cellulose and hemicelluloses and is found around 30% in conifers and broadleaves, depending on the species (Penín et al., 2020; Santos et al., 2018).

The components of smaller proportions, on the other hand, are subdivided into organic and inorganic parts and can have a great influence on the various properties of the wood. The organic part is composed of extractives, which have low and medium molecular weight and are extractable in water or neutral organic solvents (Gominho et al., 2020). These influence the color, odor, flavor, flammability, hygroscopicity and natural durability of wood (Bajpai, 2018). Included in this group are resins, essential oils, waxes, tannins and other polyphenols. In this context, it is essential to verify the compounds present in the wood, since this same wood, from urban tree pruning, will come into contact with children. The release of extractives from the toy can cause discomfort or even intoxicate the child who uses it.

2.3 Physical Properties

Wooden toys need to be durable while being lightweight and machine-friendly so that they can be shaped as needed. For pieces of wood from urban afforestation to be better disposed of, knowledge of the physical and mechanical properties of wood is essential, which avoids misuse and waste of material.

Density is one of the most important characteristics of wood, and it relates to most of its physical and technological properties (Schimleck et al., 2019; Vega et al., 2021), as well as the conditions under which the tree was planted (Costa et al., 2020). In general, heavy woods are more resistant, elastic and hard compared to light ones, but they are more difficult to work with (Barnett & Bonham, 2004). Low-density woods are also difficult to be worked on, as their fibers are pulled out when being processed. Density is a direct quantification of woody material per unit volume. It constitutes a great variability between species, between individuals of the same species and even along the trunk of a tree, since wood is a biological material that is in continuous development (Nogueira et al., 2008; Pérez-Cruzado & Rodríguez-Soalleiro, 2011). Heavy toys can be a problem for children, who can injure themselves when carrying or even playing. Thus, the ideal is to use medium-density wood for the making of toys, even facilitating the workability of the piece.

Cracking and warping in wooden toys can often occur due to the high moisture content present in the material (Arpaci et al., 2021; Odounga et al., 2018). Thus, information regarding the distribution of moisture becomes important to minimize defects. The water present inside the wood can be found in three different forms: free or capillary water, occupying partially or completely the cell lumen, the intercellular spaces and openings; impregnation or hygroscopic water, chemically bonded by hydrogen bonds, to the cellulose chains of the cell walls; and water vapor, which may be, as well as free water, contained in the empty spaces of cell cavities (Amer et al., 2019).

With the decrease of wood moisture, in addition to the loss of mass, there is a loss in volume, called volumetric shrinkage. The dimensions of the wood change substantially with the variation of moisture, in the range from 0% to the fiber saturation limit (Yuan et al., 2021). In this interval, known as hygroscopic interval, when the moisture content increases, the dimensions of the wood increase (swelling) and when the moisture content decreases, the dimensions decrease (shrinkage). However, the variation in volume in the wood occurs practically for moistures lower than approximately 28%, as the wood shows small volumetric variations, for moistures above this value (Yuan et al., 2021). This critical value for the moisture is called the fiber saturation point (FSP). Thus, it is important to properly dry the wood from urban afforestation applied to the manufacture of toys, since pieces used with high moisture can have defects over time, as they lose moisture to the environment, making them an unusable toy, or even dangerous for children.

2.4 Mechanical Properties

When subjected to external forces, the wood expresses different behaviors that determine its mechanical properties. The stress a piece of wood can withstand is affected by the direction of the load applied over the direction of fibers or tracheids, load duration, wood temperature, moisture content and density (Yuan et al., 2021). A wooden toy can be subjected to various stresses by the child who handles it, which makes it essential that the toy withstand impacts and loads to remain usable.

The wood subjected to compression stresses shows variable behavior resulting from the direction of the applied force over the direction of the fibers. It can be subjected to compression according to three orientations: perpendicular, parallel or inclined to the fibers (Namari et al., 2021). In compression parallel to the fibers, as the forces act in the same direction as the length of the wood fibers, there is great resistance. In compression perpendicular to the fibers, compaction of the fibers and elimination of voids occur, increasing the load capacity of the wood piece. Inclined compression acts both parallel and perpendicular to the fibers, being a property considered for sizing purposes (Namari et al., 2021; Yuan et al., 2021). This information helps in choosing the direction of the pieces in which the toy will be made. If the toy is liable to overload by the child who handles it, such as wooden cars, it is interesting that the pieces are made so that the load is imposed in compression perpendicular to the fibers, for example.

3 Strategies for Classification and Reuse of Urban Wood Waste

3.1 Classification and Characterization

For the correct characterization of urban wood waste, the first step to be taken is the classification of the material. Classification consists of categorizing urban wood waste, according to some specific attribute, which will allow the collected material to be divided into different classes. For urban wood waste that will be used in the production of toys, the diameter of the branches and logs can be used as the main property for the initial classification of the material, as the type of toy to be produced depends directly on the diameter of the piece to be used.

Some works have already used the classification of wood waste, according to the diameter of the branch or log, for use in the production of charcoal and in the production of toys. For charcoal production, based on carbonization yields and charcoal quality, Meira et al. (2021) only used branches with a diameter greater than 8 cm, and only 31% of the collected wood fits this diameter class, which leaves a gap for the energy reuse of branches with smaller diameters. Bispo et al. (2021), evaluating the use of wood waste from urban forests in the production of toys, used only branches larger than 15 cm, justifying that, for the satisfactory use in the manufacture of products, branches with a larger diameter are easier to work with. This opens up a research opportunity with the reuse of smaller diameter branches in the production of small wooden objects. Figure 1 illustrates a suggestion for the wood separation process, according to the diameter class, which can be performed at toy production sites.

After selecting the woods with the appropriate diameters for the production of toys, their characterizations are started. These characterizations must be carried out in laboratories with the basic structure for the correct execution of chemical, physical and mechanical analyses. In practice, public or private institutions that intend to reuse wood waste of urban origin can partner with universities or other research institutions, which would reduce analysis costs and could promote interaction between laboratory research and activities related to production of toys. Another alternative is to invest

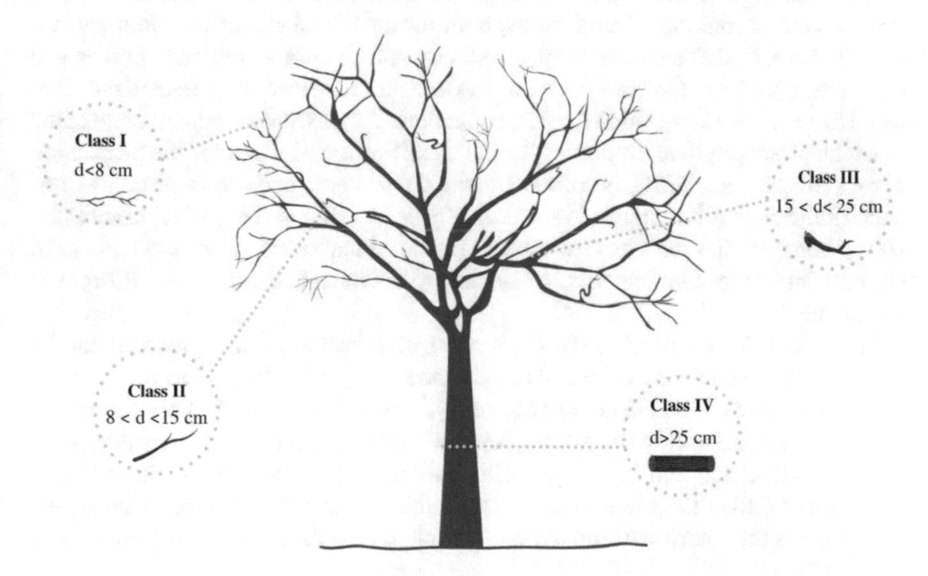

Fig. 1 Suggested division of urban wood waste classes, according to the different diameters (d)

in setting up their own laboratory, where public (city halls) or private (companies) institutions could analyze samples of material collected in the cities.

However, the creation of a laboratory is something that demands a significant amount of money, both for the purchase of equipment and automation of the site, and for the hiring and training of employees and technicians to work directly in carrying out the analyses. For this, there are already some alternatives that can facilitate the characterization of species with lower financial investments involved, in addition to the ease in classifying large quantities of wood for the production of toys. One of the simplest is the visual classification, in order to avoid the use of wood attacked by woody organisms, which may have compromised its structure, and wood with structural defects, which can compromise the process of machining and finishing the toys.

Another alternative that has been used for the characterization and grouping of wood for various purposes is colorimetry (Bispo et al., 2021; de Almeida et al., 2021; Lima et al., 2021; Martins et al., 2015). The colorimetric analysis technique allows checking the color of the wood surface through a spectrophotometer, by reading the parameters a* (color coordinate between red and green), b* (color coordinate between yellow and blue) and L* (light factor) (Sonderegger et al., 2015). The technique also allows verifying the efficiency of surface treatments applied to the wood, or even verifying the resistance of the wood against xylophagous organisms (Garcia et al., 2014; Lazarotto et al., 2016). From this, studies have also verified the feasibility of using colorimetry as a wood characterization tool, correlating the physical variables of the wood with the colorimetric patterns shown by them (Lima et al., 2021; Martins et al., 2015; Nishino et al., 2000).

With the acquisition of a portable spectrophotometer for colorimetric analysis and the correct training of employees, both for the visual characterization and for the colorimetric characterization of wood, city halls, non-governmental bodies and private companies could save financial resources in the stage of characterization of wood from urban afforestation for the production of toys. Density, which is one of the most important physical properties for the classification of wood for the production of toys (Bispo et al., 2021), is correlated with colorimetric patterns in several works, which indicate that darker woods are generally denser (Bispo et al., 2021; Lima et al., 2021). Some studies with teak wood also highlight that woods that have higher red pigment intensities and are darker have a higher content of extractives (Garcia & Marinonio, 2016).

In addition to colorimetry, there are other non-destructive analyses that can be performed to characterize the wood (Bouhamed et al., 2020; Büyüksarı & As, 2013; Niemz & Mannes, 2012; Santini et al., 2019). However, most of the non-destructive techniques are not easily accessible, as they have equipment that demand a high financial investment and high-level technical training to conduct the characterizations. These difficulties often make it impossible to use more advanced techniques, which makes the colorimetric analysis more suitable for the reality of toy production.

From the complete characterization of the woods that will be worked and the correlation with the colorimetric variables, it is possible to establish specific standards for application in the production of certain types of toys, thus making the process efficient and more economically viable.

4 Problems Arising from the Use of Urban Wood Waste and How to Get Around Them

4.1 Possible Changes in the Chemical and Physical Properties of Urban Wood Waste

Wood residue from urban forests may have some characteristics that are different from those of wood found in natural forests or planted in non-urban areas. These differences are a consequence of local environmental conditions that directly interfere with the growth of trees and, consequently, with the production of wood. The influences can be due to the space destined for the development of trees, the high exposure to atmospheric pollutants more concentrated in urban areas or even due to climate and temperature variations that can be more severe in urban environments (Meira, 2010).

As trees grow, they are able to absorb significant amounts of pollution from the environment (O_3, NO_2, PM_{10}, SO_2, CO, among others), and these different pollutants can be absorbed by leaf stomata and react with other elements inside the trees (Nowak & Dwyer, 2000). Pollutants absorbed or fixed on the surfaces of trees can be returned to nature, either by falling leaves or even by the action of rain (Nowak & Dwyer, 2000). However, this high presence of air pollutants in urban areas can significantly modify the chemical composition of wood, a factor that should be further investigated.

Trees that grow in urban areas often face problems in relation to the space available for growth and end up undergoing a series of maintenance procedures during their development (Ow et al., 2013). Whether to avoid damage to electrical wiring or the surrounding building structure, urban trees are pruned primarily for other reasons and not with the aim of enhancing the quality of the wood (Ow et al., 2013; Smiley et al., 2000). These management practices can modify the quality of wood, especially in its physical and mechanical aspects, due to the formation of knots (Ow et al., 2013). Even so, there are still few studies that assess in depth the effect of urban tree management activities on the physical and mechanical characteristics of wood. A greater understanding of these aspects is important to guide the best management strategies for these urban trees and so that different treatments that can improve the technological quality of wood waste of these species for the production of toys can be evaluated.

4.2 Possible Strategies and Solutions to Mitigate the Problems of Wood Waste from Urban Afforestation

When the collection of urban wood waste is carried out separately, that is, when it is not carried out in conjunction with other solid urban waste, a significant amount of impurities is already avoided. If the collection of the two materials is carried out together, an appropriate selection, with elimination of other residues, must be carried out. From this initial selection, a visual selection of the material can also be performed, to verify the presence or absence of defects (Bispo et al., 2021). In this initial selection, wood that has been attacked by fungi in a way that compromises the integrity and workability and branches in ways that make the workability in the production of toys unfeasible can be removed from the production of toys.

According to Bispo et al. (2021), the presence of visual defects in wood is not a limiting factor for its use in the production of toys, as these branches can be used in the production of rustic toys. Rusticity can give toys a unique look, which can be valued by consumers and, consequently, reduce the costs included with the machining and painting of wooden toys (Bispo et al., 2021). However, in addition to the visual defects present in wood, there may be defects that make its use in the production of toys unfeasible due to its low dimensional stability or high vulnerability to attack by xylophages. For cases like these, there are other types of suitable treatments, such as heat treatments.

Heat treatments are an alternative to improve the quality of wood that has some types of problems. From the action of heat on the wood, it is possible to obtain an increase in the dimensional stability of the material, mainly related to the decrease in shrinkage and the anisotropy coefficient, as well as an increase in impact flexural strength, which can contribute positively to the application of wood with unsuitable properties (Brito et al., 2006; Gaff et al., 2019). In addition, the action of heat on the wood changes the color of the surface of the material, which opens up the possibility of exploring various aesthetic aspects of urban wood waste. Changes in the chemical composition of wood species subjected to heat treatment are also observed, mainly in relation to the decrease in the extractives and hemicellulose content and, consequently, an increase in the lignin content (Corleto et al., 2020; Gaff et al., 2019).

These changes in chemical composition are important, mainly to ensure greater conservation of wood against weathering and wood-reducing organisms (Brito et al., 2008). Furthermore, thermally modified wood is a material with a promising future for use in the wooden toy industry (Ebner & Petutschnigg, 2007). Previous studies have shown that there is a significant increase in the strength of fixed joints produced with treated wood, when applied in the production of toys; in addition, treated wood is more suitable for the production of toys with moving parts (Ebner & Petutschnigg, 2007). This is one of the most promising techniques for improving the quality of wood for the production of toys.

5 Economic, Social and Environmental Benefits

Most of the biomass from urban forests in Europe and North America is used for uses with lower added value, such as land cover material or firewood for burning (Kampelmann, 2021). Although there are few studies that evaluate and quantify the use of wood from urban forests in South America, few alternatives for reusing this wood in nobler uses are known. For instance, in Brazil, the city of Rio de Janeiro, through the Municipal Urban Cleaning Company, performs an excellent job of making furniture and other wooden objects (Fig. 2) with wood waste collected in the urban forests of the city (Rio de Janeiro, 2021). These pieces of furniture and objects made by a group of street sweepers are used in the ornamentation of squares and other public environments, where the wood used to manufacture the objects came from.

In addition to making benches and other pieces of furniture that are made available for public use, handicraft items are also made using urban wood waste (Fig. 3), also creating an opportunity for an enterprise parallel to the production of furniture and other wooden objects.

Examples like this can be followed by other cities around the world, seeking the rational valorization of wood from urban forests and using them in products with direct return to communities, generating employment and income throughout the entire production chain. From this base model, others can be developed, evaluating

Fig. 2 Chairs and wooden objects made by artisan street cleaners employed by the Municipal Urban Cleaning Company in Rio de Janeiro, Brazil (*Source* William Werneck [Rio de Janeiro, 2021])

Fig. 3 Street cleaner preparing a piece of handicraft using urban wood waste (*Source* William Werneck [Rio de Janeiro, 2021])

the possibility of producing small wooden objects, such as toys, and establishing a production line dedicated to these products. Thus, it is possible to create jobs and generate income directly or indirectly for several people, directly contributing to objective 8 (decent work and economic growth) of the UN's sustainable development goals (United Nations, 2015). The creation of new jobs, even more allied with the sustainable reuse of waste, can promote different indirect benefits for the development of cities around the world. However, economic studies are still needed to assess the financial return on these investments and the size of the market that can be achieved with these activities. Some important analyses were carried out by Kampelmann (2021), verifying the parties involved in the process of using wood of urban origin for nobler purposes.

These case studies are important to serve as a basis for establishing new business models, especially knowing the opportunity available for cities to reduce their costs with the maintenance of urban forest trees. If there is a profit from the sale of wooden toys produced with wood waste from the cities themselves, collection, transport and storage costs can be significantly reduced. Other expenses are also added, considering the need for specialized labor to act in the characterization of wood and manufacture of toys.

Kampelmann (2021) suggests that an alternative for initial income generation that can finance other types of projects associated with wood from urban forests is leisure activities based on ecosystem services provided by these trees in urban areas. Thus, one of the main gaps in studies is precisely the definition of a market model and

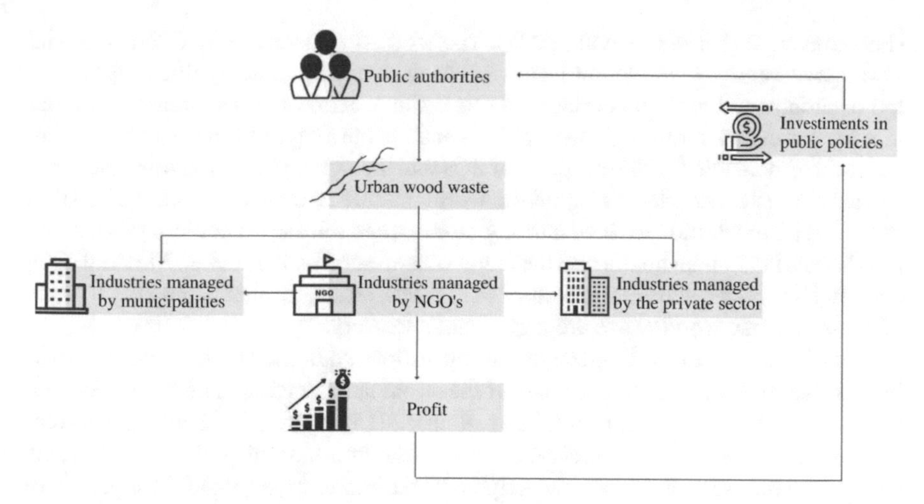

Fig. 4 Flowchart showing how the reuse of wood waste from urban sources can return as a benefit to society

the economic simulation of the implementation of this activity. Figure 4 presents a flowchart highlighting the main routes of how revenue from toy production can return in the form of benefits to the population.

Public authorities, by making urban wood waste available, can benefit directly, through the sale of wood, or indirectly, following the flow shown in Fig. 4. This role played by public authorities is essential for the next steps in the reuse of wood waste (Kampelmann, 2021). By using wood for the production of wooden toys, both the industries managed by municipalities, as well as industries managed by non-governmental organizations, and industries managed by the private sector, can profit from the products sold. From the profit obtained, more investments can be made in public policies, which will return to society through the management of resources, carried out by public authorities (Grohmann et al., 2019).

In addition to the environmental and financial benefits that this activity can generate, a social aspect, related to income, must also be taken into account: the possibility that less favored classes may have access to toys. According to research carried out by Bispo et al. (2021), on suggested prices for toy prototypes produced with urban wood waste, the values ranged between \$3.70 and \$13.58, which are relatively low prices when compared to toys sold by brands with an established world market. Lower prices allow access to wooden toys for a greater number of people, which can help in the popularization of more sustainable toys that, coupled with low cost, directly impact the reduction of environmental problems related to the incorrect disposal of this urban wood waste.

The environmental benefits of this activity should also be highlighted by stakeholders. The incorrect disposal of this material in landfills, or direct burning in the open, as it is still carried out in some cities around the world, generates a series of environmental problems related to the damage caused by the greenhouse effect

(Fetene et al., 2018; Meira et al., 2021). Therefore, the correct reuse of this material in the production of wooden objects such as wooden toys can significantly reduce the environmental problems related to this residue. Some studies already point out that the burning of biomass in the open air and the burning of biomass in homes are the main responsible for the emission of SO_2 into the atmosphere, and this practice is carried out in several places around the world, mainly in developing countries (Ren et al., 2021). In addition to SO_2, burning biomass can release other elements that are also harmful to human health and the environment, such as Particulate Material, CO, CO_2 and NO_x (Cereceda-Balic et al., 2017). Thus, strategies that enable other types of reuse of urban wood waste are urgent and necessary.

With the effective development of the suggestions addressed in this chapter, it will be possible to achieve different goals of the sustainable development goals (SDGs), adopted by the United Nations (United Nations, 2015). More specifically, the correct reuse of these wastes can significantly reduce the emission of pollutants from this source, and the creation of jobs can help boost the economy, especially for people in the lower classes, who can act both in the process of collecting and processing the urban wood waste and in the final production of toys.

6 Conclusions

The wood from urban afforestation has the potential to be used as a raw material for the production of wooden toys. With the correct selection of species, characterization of the wood and the correct training of people to work in production, it may be possible to create industries specialized in the reuse of urban wood waste. To reduce costs linked to production, partnerships with research institutions or universities can be made for the characterization of wood in laboratories. Alternative techniques, such as colorimetry, can also be important tools for wood characterization.

The establishment of the wooden toys market using wood waste of urban origin is technically feasible, but further studies are needed to verify the economic viability of the investment. Some examples of the valorization of urban wood waste already exist in European countries, but experiences with the production of toys on an industrial scale are still scarce. The main direct benefits arising from the use of urban wood waste for the production of toys can be divided into three main groups: environmental, through the reduction of polluting gas emissions due to the burning and incorrect disposal of wood; economic, as the costs involved in the management of urban trees can be deducted from the value of the sale of wood for the production of toys or even the direct sale of toys through industries managed by the municipalities; and social, with the direct generation of new jobs from the creation of this market opportunity, in addition to easier access to wooden toys, at a reduced cost, when compared to toys produced by large industries.

References

Amer, M., Kabouchi, B., Rahouti, M., Famiri, A., Fidah, A., & El Alami, S. (2019). Influence of moisture content on the axial resistance and modulus of elasticity of clonal eucalyptus wood. *Materials Today: Proceedings, 13*, 562–568. https://doi.org/10.1016/j.matpr.2019.04.014

Arpaci, S. S., Tomak, E. D., Ermeydan, M. A., & Yildirim, I. (2021). Natural weathering of sixteen wood species: Changes on surface properties. *Polymer Degradation and Stability, 183*, 109415. https://doi.org/10.1016/j.polymdegradstab.2020.109415

Bajpai, P. (2018). *Biermann's handbook of pulp and paper* (3rd ed.). Elsevier. https://doi.org/10.1016/C2017-0-00513-X

Barnett, J. R., & Bonham, V. A. (2004). Cellulose microfibril angle in the cell wall of wood fibres. *Biological Reviews, 79*(2), 461–472. https://doi.org/10.1017/S1464793103006377

Bispo, L. F. P., Nolasco, A. M., de Souza, E. C., Klingenberg, D., & Dias Júnior, A. F. (2021). Valorizing urban forestry waste through the manufacture of toys. *Waste Management, 126*, 351–359. https://doi.org/10.1016/j.wasman.2021.03.028

Bouhamed, N., Souissi, S., Marechal, P., Amar, M. B., Lenoir, O., Leger, R., & Bergeret, A. (2020). Ultrasound evaluation of the mechanical properties as an investigation tool for the wood-polymer composites including olive wood flour. *Mechanics of Materials, 148*, 103445. https://doi.org/10.1016/j.mechmat.2020.103445

Brito, J. O., Garcia, J. N., Bortoletto Júnior, G., Pessoa, A. M. das C., & Silva, P. H. M. (2006). Densidade básica e retratibilidade da madeira de Eucalyptus grandis, submetida a diferentes temperaturas de termorretificação. *Cerne, 12*(2), 182–188.

Brito, J. O., Silva, F. G., Leão, M. M., & Almeida, G. (2008). Chemical composition changes in eucalyptus and pinus woods submitted to heat treatment. *Bioresource Technology, 99*(18), 8545–8548. https://doi.org/10.1016/j.biortech.2008.03.069

Büyüksarı, Ü., & As, N. (2013). Non-destructive evaluation of beech and oak wood bent at different radii. *Composites Part B: Engineering, 48*, 106–110. https://doi.org/10.1016/j.compositesb.2012.12.006

Cereceda-Balic, F., Toledo, M., Vidal, V., Guerrero, F., Diaz-Robles, L. A., Petit-Breuilh, X., & Lapuerta, M. (2017). Emission factors for PM2.5, CO, CO2, NOx, SO2 and particle size distributions from the combustion of wood species using a new controlled combustion chamber 3CE. *Science of the Total Environment, 584–585*, 901–910. https://doi.org/10.1016/j.scitotenv.2017.01.136

Corleto, R., Gaff, M., Niemz, P., Sethy, A. K., Todaro, L., Ditommaso, G., Razaei, F., Sikora, A., Kaplan, L., Das, S., Kamboj, G., Gašparík, M., Kačík, F., & Macků, J. (2020). Effect of thermal modification on properties and milling behaviour of African padauk (Pterocarpus soyauxii Taub.) wood. *Journal of Materials Research and Technology, 9*(4), 9315–9327. https://doi.org/10.1016/j.jmrt.2020.06.018

Costa, S. E. de L., dos Santos, R. C., Vidaurre, G. B., Castro, R. V. O., Rocha, S. M. G., Carneiro, R. L., Campoe, O. C., Santos, C. P. de S., Gomes, I. R. F., Carvalho, N. F. de O., & Trugilho, P. F. (2020). The effects of contrasting environments on the basic density and mean annual increment of wood from eucalyptus clones. *Forest Ecology and Management, 458*, 117807. https://doi.org/10.1016/j.foreco.2019.117807

Cywa, K., & Wacnik, A. (2020). First representative xylological data on the exploitation of wood by early medieval woodcrafters in the Polesia region, southwestern Belarus. *Journal of Archaeological Science: Reports, 30*, 102252. https://doi.org/10.1016/j.jasrep.2020.102252

de Almeida, T. H., de Almeida, D. H., Gonçalves, D., & Lahr, F. A. R. (2021). Color variations in CIELAB coordinates for softwoods and hardwoods under the influence of artificial and natural weathering. *Journal of Building Engineering, 35*, 101965. https://doi.org/10.1016/j.jobe.2020.101965

de Morais, I. C., & Pereira, A. F. (2015). Perceived sensory characteristics of wood by consumers and trained evaluators. *Journal of Sensory Studies, 30*(6), 472–483. https://doi.org/10.1111/joss.12181

Ebner, M., & Petutschnigg, A. J. (2007). Potentials of thermally modified beech (Fagus sylvatica) wood for application in toy construction and design. *Materials and Design, 28*(6), 1753–1759. https://doi.org/10.1016/j.matdes.2006.05.015

Fetene, Y., Addis, T., Beyene, A., & Kloos, H. (2018). Valorisation of solid waste as key opportunity for green city development in the growing urban areas of the developing world. *Journal of Environmental Chemical Engineering, 6*(6), 7144–7151. https://doi.org/10.1016/j.jece.2018.11.023

Gaff, M., Kačík, F., & Gašparík, M. (2019). Impact of thermal modification on the chemical changes and impact bending strength of European oak and Norway spruce wood. *Composite Structures, 216*, 80–88. https://doi.org/10.1016/j.compstruct.2019.02.091

Garcia, R. A., & Marinonio, G. B. (2016). *Variação da Cor da Madeira de Teca em Função da Densidade e do Teor de Extrativos, 23*(1), 124–134. https://doi.org/10.1590/2179-8087.035313

Garcia, R. A., de Oliveira, N. S., do Nascimento, A. M., & de Souza, N. D. (2014). Colorimetria de madeiras dos gêneros Eucalyptus e Corymbia e sua correlação com a densidade. *CERNE, 20*(4), 509–517. https://doi.org/10.1590/01047760201420041316

Gominho, J., Lourenço, A., Marques, A. V., & Pereira, H. (2020). An extensive study on the chemical diversity of lipophilic extractives from Eucalyptus globulus wood. *Phytochemistry, 180*, 112520. https://doi.org/10.1016/j.phytochem.2020.112520

Grohmann, D., Petrucci, R., Torre, L., Micheli, M., & Menconi, M. E. (2019). Street trees' management perspectives: Reuse of Tilia sp'.s pruning waste for insulation purposes. *Urban Forestry and Urban Greening, 38*, 177–182. https://doi.org/10.1016/j.ufug.2018.12.009

Joshi, O., Grebner, D. L., & Khanal, P. N. (2015). Status of urban wood-waste and their potential use for sustainable bioenergy use in Mississippi. In *Resources, conservation and recycling* (Vol. 102, pp. 20–26). Elsevier B.V. https://doi.org/10.1016/j.resconrec.2015.06.010

Kampelmann, S. (2021). Knock on wood: Business models for urban wood could overcome financing and governance challenges faced by nature-based solutions. *Urban Forestry & Urban Greening, 62*, 127108. https://doi.org/10.1016/j.ufug.2021.127108

Lazarotto, M., Cava, S. da S., Beltrame, R., Gatto, D. A., Missio, A. L., Gomes, L. G., & Mattoso, T. R. (2016). Resistência biológica e colorimetria da madeira termorretificada de duas espécies de eucalipto. *Revista Arvore, 40*(1), 135–145. https://doi.org/10.1590/0100-67622016000100015

Lima, M. D. R., Patrício, E. P. S., Barros Junior, U. de O., Silva, R. de C. C., Bufalino, L., Numazawa, S., Hein, P. R. G., & Protásio, T. de P. (2021). Colorimetry as a criterion for segregation of logging wastes from sustainable forest management in the Brazilian Amazon for bioenergy. *Renewable Energy, 163*, 792–806. https://doi.org/10.1016/j.renene.2020.08.078

Lousada, C. M. (2020). Wood cellulose as a hydrogen storage material. *International Journal of Hydrogen Energy, 45*(29), 14907–14914. https://doi.org/10.1016/j.ijhydene.2020.03.229

Martins, M. D. F., Beltrame, R., Delucis, R. D. A., Gatto, D. A., De Cademartori, P. H. G., & Dos Santos, G. A. (2015). Colorimetria como ferramenta de agrupamento de madeira de clones de eucalipto. *Pesquisa Florestal Brasileira, 35*(84), 443. https://doi.org/10.4336/2015.pfb.35.84.929

Meira, A. M. (2010). *Gestão de resíduos da arborização urbana* [Biblioteca Digital de Teses e Dissertações da Universidade de São Paulo]. https://doi.org/10.11606/T.11.2010.tde-19042010-103157

Meira, A. M., Nolasco, A. M., Klingenberg, D., Souza, E. C., & Dias Júnior, A. F. (2021). Insights into the reuse of urban forestry wood waste for charcoal production. *Clean Technologies and Environmental Policy*. https://doi.org/10.1007/s10098-021-02181-1

Migneault, S., Koubaa, A., Perré, P., & Riedl, B. (2015). Effects of wood fiber surface chemistry on strength of wood–plastic composites. *Applied Surface Science, 343*, 11–18. https://doi.org/10.1016/j.apsusc.2015.03.010

Morata, A., González, C., Tesfaye, W., Loira, I., & Suárez-Lepe, J. A. (2019). Maceration and fermentation. In *Red wine technology* (pp. 35–49). Elsevier. https://doi.org/10.1016/B978-0-12-814399-5.00003-7

Murphy, E. K., Mottiar, Y., Soolanayakanahally, R. Y., & Mansfield, S. D. (2021). Variations in cell wall traits impact saccharification potential of Salix famelica and Salix eriocephala. *Biomass and Bioenergy, 148*, 106051. https://doi.org/10.1016/j.biombioe.2021.106051

Namari, S., Drosky, L., Pudlitz, B., Haller, P., Sotayo, A., Bradley, D., Mehra, S., O'Ceallaigh, C., Harte, A. M., El-Houjeyri, I., Oudjene, M., & Guan, Z. (2021). Mechanical properties of compressed wood. *Construction and Building Materials, 301*, 124269. https://doi.org/10.1016/j.conbuildmat.2021.124269

Niemz, P., & Mannes, D. (2012). Non-destructive testing of wood and wood-based materials. *Journal of Cultural Heritage, 13*(3), S26–S34. https://doi.org/10.1016/j.culher.2012.04.001

Nishino, Y., Janin, G., Yainada, Y., & Kitano, D. (2000). Relations between the colorimetric values and densities of sapwood. *Journal of Wood Science, 46*(4), 267–272. https://doi.org/10.1007/BF00766215

Nogueira, E. M., Fearnside, P. M., & Nelson, B. W. (2008). Normalization of wood density in biomass estimates of Amazon forests. *Forest Ecology and Management, 256*(5), 990–996. https://doi.org/10.1016/j.foreco.2008.06.001

Nowak, D. J., & Dwyer, J. F. (2000). Understanding the benefits and costs of urban forest ecosystems. In *Handbook of urban and community forestry in the northeast* (pp. 11–25). Springer US. https://doi.org/10.1007/978-1-4615-4191-2_2

Nowak, D. J., & Greenfield, E. J. (2018). US urban forest statistics, values, and projections. *Journal of Forestry, 116*(2), 164–177. https://doi.org/10.1093/jofore/fvx004

Nowak, D. J., Greenfield, E. J., & Ash, R. M. (2019). Annual biomass loss and potential value of urban tree waste in the United States. *Urban Forestry and Urban Greening, 46*, 126469. https://doi.org/10.1016/j.ufug.2019.126469

Odounga, B., Moutou Pitti, R., Toussaint, E., & Grédiac, M. (2018). Mode I fracture of tropical woods using grid method. *Theoretical and Applied Fracture Mechanics, 95*, 1–17. https://doi.org/10.1016/j.tafmec.2018.02.006

Ow, L. F., Ghosh, S., & Sim, E. K. (2013). Mechanical injury and occlusion: An urban, tropical perspective. *Urban Forestry & Urban Greening, 12*(2), 255–261. https://doi.org/10.1016/j.ufug.2013.02.004

Penín, L., López, M., Santos, V., Alonso, J. L., & Parajó, J. C. (2020). Technologies for eucalyptus wood processing in the scope of biorefineries: A comprehensive review. *Bioresource Technology, 311*, 123528. https://doi.org/10.1016/j.biortech.2020.123528

Pérez-Cruzado, C., & Rodríguez-Soalleiro, R. (2011). Improvement in accuracy of aboveground biomass estimation in Eucalyptus nitens plantations: Effect of bole sampling intensity and explanatory variables. *Forest Ecology and Management, 261*(11), 2016–2028. https://doi.org/10.1016/j.foreco.2011.02.028

Pournou, A. (2020). Wood deterioration by insects. In *Biodeterioration of wooden cultural heritage* (pp. 425–526). Springer International Publishing. https://doi.org/10.1007/978-3-030-46504-9_7

Ren, Y., Shen, G., Shen, H., Zhong, Q., Xu, H., Meng, W., Zhang, W., Yu, X., Yun, X., Luo, Z., Chen, Y., Li, B., Cheng, H., Zhu, D., & Tao, S. (2021). Contributions of biomass burning to global and regional SO2 emissions. *Atmospheric Research, 260*, 105709. https://doi.org/10.1016/j.atmosres.2021.105709

Rio de Janeiro. (2021). *Garis artesãos da Comlurb transformam troncos de árvores em arte - Prefeitura da Cidade do Rio de Janeiro - prefeitura.rio.* https://prefeitura.rio/comlurb/garis-artesaos-da-comlurb-transformam-troncos-de-arvores-em-arte/

Santini, L., Jr., Ortega Rodriguez, D. R., Quintilhan, M. T., Brazolin, S., & Tommasiello Filho, M. (2019). Evidence to wood biodeterioration of tropical species revealed by non-destructive techniques. *Science of the Total Environment, 672*, 357–369. https://doi.org/10.1016/j.scitotenv.2019.03.429

Santos, T. M., Alonso, M. V., Oliet, M., Domínguez, J. C., Rigual, V., & Rodriguez, F. (2018). Effect of autohydrolysis on Pinus radiata wood for hemicellulose extraction. *Carbohydrate Polymers, 194*, 285–293. https://doi.org/10.1016/j.carbpol.2018.04.010

Schimleck, L., Dahlen, J., Apiolaza, L. A., Downes, G., Emms, G., Evans, R., Moore, J., Pâques, L., Van den Bulcke, J., & Wang, X. (2019). Non-destructive evaluation techniques and what they tell us about wood property variation. *Forests, 10*(9), 728. https://doi.org/10.3390/f10090728

Smiley, E. T., Fraedrich, B. R., & Fengler, P. H. (2000). Hazard tree inspection, evaluation, and management. In *Handbook of urban and community forestry in the northeast* (pp. 243–260). Springer US. https://doi.org/10.1007/978-1-4615-4191-2_17

Sonderegger, W., Kránitz, K., Bues, C. T., & Niemz, P. (2015). Aging effects on physical and mechanical properties of spruce, fir and oak wood. *Journal of Cultural Heritage, 16*(6), 883–889. https://doi.org/10.1016/j.culher.2015.02.002

United Nations. (2015). *Agenda of sustainable development goals 2030*. https://sdgs.un.org/goals

Vega, M., Harrison, P., Hamilton, M., Musk, R., Adams, P., & Potts, B. (2021). Modelling wood property variation among Tasmanian Eucalyptus nitens plantations. *Forest Ecology and Management, 491*, 119203. https://doi.org/10.1016/j.foreco.2021.119203

Yuan, Q., Liu, Z., Zheng, K., & Ma, C. (2021). Wood. In *Civil engineering materials* (pp. 239–259). Elsevier. https://doi.org/10.1016/B978-0-12-822865-4.00005-2

Zabel, R. A., & Morrell, J. J. (2020). *Wood microbiology* (2nd ed.). Elsevier. https://doi.org/10.1016/C2018-0-05117-8

Zhang, C., Mo, J., Fu, Q., Liu, Y., Wang, S., & Nie, S. (2021). Wood-cellulose-fiber-based functional materials for triboelectric nanogenerators. *Nano Energy, 81*, 105637. https://doi.org/10.1016/j.nanoen.2020.105637

The Pile of Shame: The Personal and Social Sustainability of Collecting and Hoarding Miniatures

Mikko Meriläinen, Jaakko Stenros, and Katriina Heljakka

Abstract Collecting is a major part of the miniaturing pastime, in which enthusiasts collect, paint, and play games with small historical and fantasy wargaming and role-playing figurines. Miniaturists often have large collections of miniatures, and many buy more miniatures than they have time to paint. This quantity of unpainted miniatures is often referred to as a *pile of shame.* In this chapter, we explore the collecting of miniatures and the pile of shame phenomenon through a thematic analysis of qualitative survey data ($N = 127$). Our analysis suggests that an amassed collection of miniatures poses both practical and existential potential and challenges and may be both beneficial and detrimental to personal sustainability. Although the concept of a pile of shame is typically a shared source of humour, it is also a relevant part of the miniaturing pastime, and an important aspect of how miniaturists curate and view their collection.

Keywords Miniaturing · Wargaming · Collecting · Hoarding · Adult play · Warhammer · Consumer behaviour

1 Introduction

> *For my favourite army I hoard everything though. It's some weird thing, you have to have too much and even more.* (ID 125)

M. Meriläinen (✉) · J. Stenros
Game Research Lab, Tampere University, Kanslerinrinne 1, 33014 Tampere, Finland
e-mail: mikko.merilainen@tuni.fi

J. Stenros
e-mail: jaakko.stenros@tuni.fi

K. Heljakka
Degree Programme of Cultural Production and Landscape Studies, University of Turku, Siltapuistokatu 11, 28100 Pori, Finland
e-mail: katriina.heljakka@utu.fi

S. S. Muthu (eds.), *Toys and Sustainability*, Environmental Footprints and Eco-design of Products and Processes,
https://doi.org/10.1007/978-981-16-9673-2_4

Collecting is a major part of *miniaturing*, the pastime of engaging with small fantasy wargaming and role-playing figurines that lies at the intersection of games, toys, play, and crafting (Meriläinen et al., 2020). Miniaturing enthusiasts commonly amass collections of hundreds or even thousands of miniatures, often buying miniatures faster than they can paint them. As a result, the topic of the *pile of shame* or *lead mountain*, referring to significant amounts of unpainted miniatures, is a common source of both humour and anguish among enthusiasts, with references to hoarding behaviour. This is only a part of the picture, however, and there is a multitude of buying and collecting behaviours directed by personal preferences and finances as well as external influences.

This widespread practice of accumulating miniatures has interesting implications for personal, social, and environmental sustainability. The miniaturing pastime pivots around the material miniature figurine (Meriläinen et al., in press). They are bought, stored, assembled, painted, displayed, and toyed and played with. In the circular economy of miniaturing, there is little waste resulting from the activity: unused miniatures and miniature components are treasured, and miniature builders often re-purpose waste materials, such as soda cans, plastic containers, and scraps of cardboard packaging materials in building figurines or dioramas.[1] Yet it is not uncommon to have piles of unpainted miniatures, possibly from numerous decades. There is a significant monetary cost involved, and this activity, which can look like hoarding, requires negotiation in family relationships, for example, in terms of investments in finances, time, and space, but also bargaining with oneself to justify the decisions related to buying, trading, and selling miniatures. Recently, digital miniatures and 3D printing have further complicated the issue of collecting and hoarding.

In this qualitative study, we examine miniature enthusiasts' collecting behaviours and the views associated with them with a thematic analysis (Braun & Clarke, 2006, 2012) of rich online qualitative survey data from 127 Finnish adult miniaturing enthusiasts. Our self-selected group of respondents mainly situate themselves in the post-*Dungeons & Dragons* fantasy wargaming and role-playing scenes, with Games Workshop's *Warhammer* as the most popular, though not only, commercial franchise. This data is used as a foundation for a discussion of miniature enthusiasts and their accumulation of physical figurines.

This chapter first explores the background of the miniaturing pastime and its connection to collecting and consumption. This is followed by a discussion of the research method and data. We then move on to analyse the practical and existential dimensions of collecting miniatures. In the discussion and conclusions, we consider the life cycle of a miniature and the phenomenon of the pile of shame as a starting point for negotiating personal sustainability of the pastime and one's collection as a personal imagination.

---

[1] Diorama, in the context of fantasy and historical miniatures, refers to a three-dimensional scale model of a situation, containing miniatures in a scene, such as a fantasy battle or a moment on an adventure.

2 Background

Wargaming is a practice that has a long and complex history drawing on toys and games and stretching back to tin soldiers, Prussian military simulations, and *chess* (e.g. Lewin, 2012; Peterson, 2012). Our study, however, concentrates on contemporary miniaturists. These miniature enthusiasts have received relatively little attention in game studies (e.g. Carter, Gibbs, et al., 2014; Carter, Harrop, et al., 2014; Cova et al., 2007; Harrop et al., 2013; Kankainen, 2016; Meriläinen et al., 2020, in press; Williams & Tobin, 2021). In this background section, we discuss miniatures and miniaturing especially in the context of toys, the "pile of shame" phenomenon in miniaturing and its comparable manifestations in other craft and play cultures and discuss the consumption and sustainability of miniatures.

2.1 Miniatures and Miniaturing

In this chapter, we define miniatures as scaled-down plastic or metal representations of historical and fictional characters, creatures, and objects, typically used in wargaming, role-playing games, and for display purposes. Miniatures are typically either single-part castings or prints, or otherwise non-poseable after construction from parts. There are different sizes of miniatures available, usually using the height of a typical humanoid character in millimetres for reference. Common sizes used for gaming are 10 mm, 15 mm, and especially the loosely defined 28–32 mm bracket, while larger sizes such as 54 mm and 75 mm typically find more use as display pieces.

A miniaturist is someone who engages in the miniaturing pastime (see Meriläinen et al., 2020) by, for example, painting miniatures, playing games with them, or discussing them with other people. It is important to note that we use the word "miniaturist" as a broad, inclusive descriptor. Although it is likely also a social identity for some, we have elected for this interpretation as the topic has not yet been studied sufficiently. The word covers a broad selection of ways to engage with miniatures, from passionate collectors to tournament players and from professional painters to role-playing gamers for whom miniatures mainly serve as gaming tokens.

While miniatures function as game pieces and collectibles, they are also toys. Traditionally, toys are considered a material, tactile, and narrative medium and as such miniatures are part of the wider material cultures of play. Their primary affordance is to be playable—physically, spatially, and narratively manipulatable objects—both as toys and as game pieces. Furthermore, as vintage toy researcher Jonathon Lundy (2021, p. 195) has noted, "[t]he materiality of toys is also at the root of their collectability". In adulthood, toys are often employed as part of collecting practices, but also in the performative acts of curation and creative cultivation. Toy collecting in adulthood can be coloured by nostalgia, with key valuations born in childhood, but in essence, this practice is more complex than sentimental. For some toys, collectability is deeply embedded in both the plaything and the logic and mechanisms of play characterizing these objects.

The longevity of miniatures as physical objects allows individuals to track down and acquire miniatures from their youth or childhood, although often at inflated prices; with time these originally relatively cheap gaming pieces have become valuable out of production collectibles. With older miniatures produced in the 1970s and 1980s, this is not a case of artificial scarcity, as the rubber moulds used to cast miniatures eventually decay with use and time, master castings and original sculpts are destroyed in the moulding process or simply lost, making it impossible to reproduce old miniatures in their original form.

What differentiates engagement with miniatures from playing with mass-produced toys, such as action figures, is the centrality of the crafting aspect (see Meriläinen et al., 2020): most miniatures are provided unpainted and, in some cases, unassembled, and there is an implicit, or even explicit, notion that they will be painted. Pre-painted miniatures are an exception to this, as they are marketed as both collectibles and as gaming pieces. Even with pre-painted miniatures, however, the crafting dimension is present. For example, WizKids, a company that produces miniatures compatible with the popular *Dungeons & Dragons* tabletop role-playing game, produces both unpainted and painted versions of their miniatures. The pre-painted ones are labelled "premium"—implying that the work required to paint the miniatures has already been done and the models are ready to be used (WizKids, 2021).

Miniatures often form parts of transmedial complexes, either in a central or in a peripheral role. An example of centrality is the miniatures of Games Workshop's *Warhammer* franchise and its various extensions. Here, a transmedial world consisting of digital games, books, comics, films, and other toys has been constructed around miniatures and the tabletop games played with them. Miniatures can also be peripheral, as in the above case of *Dungeons & Dragons*, in which miniatures can be used to support gameplay and to visualize imaginary characters, or when miniatures are either officially or unofficially created from existing intellectual property, such as Micro Art Studios' miniatures depicting the characters of the popular *Discworld* series of fantasy books.

2.2 The Pile of Shame

Miniaturing is very material: the diverse activities miniaturists engage in as part of the pastime all revolve around the physical miniatures to some extent (Meriläinen et al., in press). Miniatures are acquired for different purposes, such as for painting, gaming, displaying, toying, and collecting, and for most miniaturists their engagement with the pastime features several of these dimensions. As a result, many miniaturists amass considerable collections of miniatures consisting of hundreds or even thousands of miniatures, which sometimes poses challenges in terms of use of time, storage, and personal finances.

In adult toy cultures, toys are sometimes referred to as "plastic crack" (see Lundy, 2021). Collecting toys costs money and while they are not addictive per se, collecting is a strong play pattern across generations of toy enthusiasts, and it is certainly encouraged by the toy industry and many toy cultures. In miniaturists' vernacular, the amassed collection of unpainted miniatures is often affectionately referred to as a "pile of shame". Despite the use of the word "shame", the pile of shame appears more as an in-joke, a convenient shorthand for the fact that it is often much quicker to acquire than to paint miniatures. The topic of an excess of unpainted miniatures is a common source of humour in miniaturist communities: it appears often in memes, and Games Workshop, the world's current leading miniatures manufacturer, even published a video titled *Fifty Shelves of Grey*—satirizing the popular *Fifty Shades of Grey* novels and movies and referring to shelves full of the miniatures they produce, which are grey plastic in their unpainted state.

Miniatures are obviously not the only adult toys that people amass enough to feel bad about, and the pile of shame is not an expression limited to miniaturing, or even hobbies more generally. Regardless of whether it refers to discount clothes (Shell, 2009), digital games (Johnson & Luo, 2019), or tabletop games in the form of the *shelf of shame* (Coward-Gibbs, 2021), the meaning is typically the same. Interesting parallel practices related to the thinking of the miniature collection as a material resource offering affordance for both actual crafting and the imagination could likely be found in other craft communities. For example, buying more yarn than one can ever knit seems like an interesting point of comparison.

We suggest here that the pile of shame is the collection of things one has acquired and not yet used—and it weighs particularly heavily when one acquires new things even if old ones are still waiting to be used. Owning something for the sake of collecting, for example, is not seen as a viable use: a miniature should be painted, or a board game should be played.

Collecting is an important part of miniaturing, but it appears that for most miniaturists collecting also implies using miniatures in other ways besides hunting them down in the marketplace, owning or storing them. As Williams and Tobin (2021) note in their article on Oldhammer, the miniaturist subculture centred on retro Warhammer miniatures: "To Oldhammer is to be active, to craft and to play with things, not just appreciate them; Oldhammer is not really antiquarian". Seen against this backdrop, the idea of a pile of shame is understandable—as long as a miniature remains unpainted, it has not seen "proper" use.

In recent years some miniaturists have discussed moving away from the framing around shame and considering an alternate framing as the *pile of potential* (see davekay, 2020; Wudugast, 2020; see also Coward-Gibbs, 2021). This approach seeks to dissolve or remove the dimension of shame, even if joking, from the equation, and encourages miniaturists to enjoy their enthusiasm even if not every project gets completed. Attention is drawn to the miniatures' potential: although tastes may change and interest may wane, the miniature still offers potential enjoyment in the future.

2.3 Sustainability and Consumption

The sustainability of the miniaturing pastime has not been studied previously. In this chapter, we concentrate on the personal and social sustainability of the activity (e.g. Dhar et al., 2021). From the point of view of sustainability and consumption, toys are a complicated subject. While play in general is widespread in the animal kingdom, and quite a few species even play with objects, humans are the only species that specifically makes toys, objects for play (Burghardt, 2005). Toys need not be useful in goal oriented activities, but playful in activities that are meaningful in themselves. It is difficult to assess their sustainability from a usefulness or efficiency point of view. Yet in general, toys offer a multitude of possibilities for skill-building and cultivating personal creativity through physical manipulation and narrative meaning making, which makes toys an interesting case to study topics such as adult imagination and crafting.

That said, most contemporary toys are no longer made by the humans who use them or their parents. They are commodities, manufactured for consumption. According to play scholar Brian Sutton-Smith (2017, pp. 233–234), this goes even further, from toys to play. According to him the most obvious modern manifestations of play appear as *consumable experiences*—for example, as toys and computer and video games. So-called *commoditoys* (Langer, 1989) stimulate consumption by design.

Hannah Arendt (1968/2007) has commented that collecting is a passion of children (things need not be valued as commodities or seen as useful)—and a hobby of the rich who do not need things to be useful and can afford to make "transfigurations of objects". While collecting today can be an investment as well (see Belk et al., 1991), for the most part the collectors can dream with their items in a way that "things are liberated from the drudgery of usefulness". This points towards collecting, in itself, as play (see also Heljakka, 2013). Just as a bibliophile may not read their collected books (Benjamin, 1931/2007) or a digital gamer may not play the games they've bought (Johnson & Luo, 2019), a miniature enthusiast may not need to actually play with their figurines or even assemble them.

Assessing toys from the point of view of usefulness is complicated. Can owning be "use"? What form does this "use" take? Is a miniature "done" when its crafting has finished and it is being displayed, or does it continue life as a gaming piece and a pivot of fantasy and imagination, drawing from transmedia universes, play experience, and personal meaning making? These are some of the questions we address in light of the data collected. Miniatures invite long-term play behaviour—engagement with their materiality and multiplicity. Toys and toy collections have an existential aspect as well, functioning as mirrors for the collectors and potentially outlasting them. Miniature hobbyists struggle with the dilemma between being buried with one's toys and being buried in toys.

3 Method and Data

In this study, we explore miniaturing and the pile of shame phenomenon utilizing a qualitative data set ($N = 127$) collected in 2019 with a Finnish language online questionnaire. The questionnaire consisted of seven open questions and six demographic questions and was distributed online on Facebook groups, hobby forums, and Twitter.

Most of the self-selected respondents (91.3%, $N = 116$) identified as men, with only 7.9% of respondents ($N = 10$) identifying as women and a single participant (0.8%) not disclosing gender information. This appears to reflect the gendered nature of the miniaturing hobby (see Körner & Schütz, 2021; Singleton, 2021). Our participants were between 18 and 56 years of age, with a median of 35. As the questionnaire was aimed at adult miniaturists, the minimum age for participation was 18. Our participants can be described as experienced miniaturists, as the median year for starting miniaturing was 1998, and the earliest one 1970. This is a relevant feature of our sample, as it means that many of our respondents have had time to amass considerable collections of miniatures.

To explore our data, we conducted a thematic analysis. Thematic analysis is a qualitative research method, in which researchers identify both novel items and recurring patterns in a set of data and organize these observations into broader themes (Braun & Clarke, 2006). Our thematic analysis was a semantic one; we focused on what was actually written by the respondents. We had previously conducted two thematic analyses on the data (Meriläinen et al., 2020, in press). These two analyses, one a general exploration of the data and the other focusing on the materiality of miniatures, served as a starting point which we then supplemented with a new round of analysis. Instead of utilizing a pre-existing theoretical frame, we conducted an inductive, exploratory analysis, focusing on responses discussing either the pile of shame phenomenon directly or addressing issues such as the collecting and storing of miniatures.

The coding was conducted using the *Atlas.ti* software, a qualitative data analysis tool. After coding, the codes were organized first into smaller subthemes which were then combined to form two main themes that provide distinct points of view into the collecting of miniatures: *Practical sustainability* and *Existential sustainability*.

4 Results

In our analysis, we identified two distinct dimensions related to how miniaturists view their miniature collections and their acquisitions of miniatures. *Practical sustainability* addresses miniaturing as a situated activity. In the theme we explore miniatures as material objects that occupy space, demand time, and have monetary value, and the everyday concerns and considerations related to these aspects. *Existential sustainability* examines the personal meanings miniaturists assign to miniatures and miniaturing. Interestingly, while many of the responses detailed the "pile of shame" phenomenon, only a single individual explicitly discussed it in their response with that name.

We have presented quotes from the data to illustrate our themes. The quotes have been translated from Finnish, and minor corrections such as capitalization of sentences have been made. However, we have sought to preserve as much of the form and the tone of the quotes as possible.

4.1 Practical Sustainability

Although the pile of shame is an abstract notion and can mean anything from tens to thousands of unpainted miniatures, it is not imaginary. While miniatures may not literally be stored in piles, they take up physical space whether stored in glass cabinets or storage boxes (see Meriläinen et al., in press). They need to be bought with money or acquired through trades or as gifts, and time must be allocated for painting them. Space, time, and money are all limited resources, and miniaturists need to negotiate all three to keep their pastime sustainable. The theme illustrates miniaturing as a situated activity (see Apperley, 2010): miniaturing takes place in the confines of everyday life and is influenced and regulated by many things other than the miniaturist's personal wants, needs, and preferences.

Practical considerations are tied to place and time. Many of our respondents considered both their current and former practices, sometimes stretching over several decades. With changes in living conditions, life situations, and available finances, their miniaturing practices had also changed. Many respondents mentioned that they had started to consider their purchases more carefully compared to a less restricted and reflective past, as they had already amassed what they felt was enough miniatures for the time being. Informants reflected on their resources, both monetary and miniature materials. As shown in the quotes below, many informants regulated their spending, bargained with themselves, and set up personal systems and rules for miniature purchasing and crafting.

> I nowadays try to avoid buying big expensive new sets. I think about recycling a lot - can I mod [modify] something for a new purpose? Can I take apart a vignette and how can I use that? My buying behaviour has changed - when I was younger I wouldn't have given any extra thought to it but would've just thrown all of my money into some new expensive set because it's nice without any thought to whether it was necessary. Now I think more about use and recycling. If there's something I have to have, I ask my friends whether anyone knows about second-hand availability, I can't bring myself to buy almost anything new. (ID 7)

> Nowadays I don't buy [miniatures] at all, because I have tens of unpainted ones left from my "wild years". [...] My most intensive buying period was probably as a 14–25-yearold, when I was more of a hoarding type anyway. (ID 13)

Views on the use of money differed based on personal priorities and available finances. Many of our respondents had stopped miniaturing at some point because they felt it was too expensive. Referring especially to Games Workshop and their increasing prices was common, reflecting anti-corporate sentiments expressed by a segment of miniaturists (Williams & Tobin, 2021). A recurring narrative in the data was that of increasing personal finances providing more affordances for miniaturing. Although actual sums used on miniaturing varied, many respondents explicitly mentioned that their buying behaviour was not excessive compared to their available money. Some were actively keeping track of their spending or had an allocated budget for miniatures. Spending money was also considered in relation to available time: some respondents had very limited time available, so they went for a "quality over quantity" approach.

> If I compare my [current] buying behaviour to the early 90s it's like day and night. As a student, GW's [Games Workshop's] new price changes practically made the hobby so expensive that it was a major reason for a 20 year break. Nowadays even though the prices are high, they're not as considerable in relation to my income as a child/young adult. (ID 17)

> Oh boy. My favourite topic. It's because I have kept track of all of my spending on miniatures and associated activities as carefully as possible, and I can tell you that in almost 17 years the accumulated sum is horrifying. I also relate money spent on a given game system to the amount of gaming sessions, so I can calculate a mean for every game played. When the sum is under a certain amount, I am "permitted" to buy miniatures for said game. (ID 91)

> I have realized that in this life situation time is the resource used to pay for things. I'm happy to pay several tens of euros for an especially nice mini that I know I'll have many amazing painting sessions with and eventually finish, whereas even the idea of painting a hundred orcs, even if I got them for free, horrifies me. (ID 119)

Excess miniatures, whether painted or unpainted, were sold off for a variety of reasons, such as lack of space, interest, or time. Selling a project could also provide a sense of completion and closure: owning the miniatures was sometimes seen as secondary to working on them.

> I don't care much about selling them, mainly I sell off miniatures I no longer need. I feel like selling them takes too much time and effort. But sometimes I also have to sell things, a person living in the city, in an apartment does not have too much storage space. (ID 35)

> I also sell off finished armies as used ones (as well as unopened projects…), so maybe it's more about finishing projects rather than collecting. (ID 51)

> Being middle aged I've started selling off some of my miniatures, the idea being "if I haven't managed to paint this in 20 years then maybe it's time for it to go to another owner". (ID 53)

Not all miniatures were viewed as equal, and extremely relevant to considerations of the pile of shame phenomenon was the idea of miniatures becoming surplus at some point. There were a variety of reasons, sometimes overlapping, for surplus miniatures: some were the results of shopping sprees or sales, others resulted from stalled projects, waning interest, or changes in game preferences. At the other end of the spectrum were very specifically purchased models, whether old out of production pieces, specific representations of role-playing game characters, or collectibles. When getting rid of miniatures, it was the surplus that was cut.

> We [refers to respondent and their partner] haven't sold a single mini, and because of sentimental reasons the minis of old player characters will probably never be sold. (ID 88)

> Special offers in online stores often bring about temptation. If I buy, I concentrate my buying and buy a little bigger lot at once. I do sell off miniatures I no longer use or ones that ended up being wasted purchases. (ID 31)

As collections of miniatures cost money and take up domestic space as well as time, some respondents had to negotiate their hobby with their partner. Although most partners appeared to have a positive view of miniaturing or even participated in it themselves, space sometimes became a point of contention. Any conflict mentioned appeared to be minor, however.

> The wife is also a miniature hobbyist, before [having] kids we gamed and painted together approx. on a weekly basis, that has dropped off now. We dream that when the kids are older, we'll get it back. (ID 63)

> The minis are in a large cabinet, not especially on display but not hidden either. My wife doesn't want there to be tanks or Roman cavalry on the bookshelf. I accept this. [...] My wife's statement is: "I accept but I don't understand". (ID 92)

> My partner doesn't quite understand my hobby and feels that I own too much stuff related to it. It annoys them that I have many boxes full of miniatures, but they nevertheless support me in my hobby, they just wish there wasn't so much stuff. It is because of this that I keep my minis hidden in the basement and advertise it when I finish something or sell it off (meaning that there is less stuff). (ID 130)

As the examples above show, the miniature collection prompts several practical and social considerations apart from any additional meaning attributed to it. Regardless of the miniaturist's personal views, the availability of space, time, and money can all place very concrete limitations and provide affordances for the pastime. From the point of view of these practical factors, the sustainability of miniaturing lies in balancing available time, space, and money. Ideally, the miniaturist can afford those miniatures they want, have space for the miniatures they have and intend to purchase, and have time to enjoy their miniatures. This practical dimension, however, is only a part of the whole. It is the less tangible but extremely relevant personal meanings attributed to miniatures and miniaturing that we turn to next.

4.2 Existential Sustainability

The importance and worth of miniatures are often detached from their monetary value and other practical considerations. This theme addresses miniatures and miniaturing as something given personal value. Hence, it deals with *personal sustainability* (e.g. Dhar et al., 2021) and how collecting miniatures is tied to different pleasurable activities and experiences that help reduce stress, drive everyday renewal, and improve well-being.

As discussed above and in previous research (Carter, Gibbs, et al., 2014; Meriläinen et al., 2020, in press; Williams & Tobin, 2021), miniaturing consists of multiple activities such as painting miniatures, playing games with them, socializing with friends, and displaying miniatures at home or online. For miniaturists, these activities can be psychologically very rewarding. Reducing stress is a key element of personal sustainability (Dhar et al., 2021), and many of our respondents explicitly posited miniaturing as a counterbalance to the demands of other life areas, typically work but also family life. Miniaturing provided different ways to reduce stress, often formed around immersion either in the activity itself, the fiction related to it, or both. Moreover, miniaturing provided avenues for creativity and self-expression not just through the crafting aspect, but also through tactical gaming and prompting storytelling, as illustrated by the quotes below.

> It is a sandbox of creative activity where I can move those elements of imagination and tactical ambition that I find it difficult to find a space for in the rest of my life. (ID 22)

> A large part of it is distancing yourself from everyday life and other things. Painting, assembling, and planning take much more time than the gaming itself and you feel quite "zen" when you can immerse yourself in the painting of a new miniature - or just planning the colour scheme with a pen and paper. It keeps the mind lively;). (ID 72)

> Painting is often a very relaxing, meditative experience for me. I usually listen to music or podcasts while doing it. I also like problem-solving related to painting and assembling: how do I paint a glowing torch? What about pallid skin or rusty armour? Solving these small problems provides me with experiences of success. The nature of my day job is that nothing ever really gets finished, there are no projects or deadlines. Finishing miniature projects provides satisfaction compared to that. (ID 124)

As the above quotations show, miniaturing is a controlled province of meaning for the informants; it is something that they feel mastery over, something that is their own, and where they can let their imagination run free. Miniaturing allows miniaturists to learn new things as need arises, take on limited projects and succeed in carrying them out, and set their own pace. Miniaturing, especially the crafting aspect of it, is about shaping material figurines and the fantasy around them and exerting calm and control over them.

In this chapter, the focus is on collecting miniatures. However, the context provided by the quotes above is crucial to understanding why people accumulate miniatures: they are often acquired for their promise of pleasant activities such as painting and gaming. As one of our respondents summarizes it:

> Every serious hobbyist has the same problem with a lead/plastic mountain, but that is perhaps more due to unrealistic ideas about future projects than pure shopping euphoria. (ID 32)

The topic of projects and planning is prevalent in the data. Many of our respondents described their miniaturing through different projects, whether these were entire armies, units, warbands, or individual player characters. Miniatures were acquired to fit these projects, something that the respondents often framed as *having a use for* or *needing* something, and posited as an opposite to hoarding, or buying miniatures for the sake of having more miniatures. The surplus miniatures mentioned in the previous theme were typically miniatures that no longer had this type of use or had never had it to begin with.

> I only buy those [miniatures] that I have some kind of use for, even though I'd like to buy this and that just for the joy of painting. (ID 12)

> I don't necessarily think about the price of something if I really want it. However, I don't blindly buy whatever I run into, but it has to have at least some use, even if it's just cutting it up for conversion [miniature customization] parts. (ID 70)

> A while back I bought quite a lot of minis in one go, I have enough minis in the cupboard to finish 3 large armies and for several smaller projects. I have now intentionally limited my buying, I try to finish minis I've already bought first. In that sense I try to keep my miniaturing hobby sustainable in that sense, so I try to avoid hoarding. However, in my head there are constantly future projects and all the things I should get for them. In my gaming group most appear to be hoarders, in other words they buy much more than they have time to paint. (ID 124)

Respondents reflected on their buying behaviour, relating it to their life and well-being. Even though many spent considerable sums on miniatures, this was seen as an investment into personal sustainability. While they were spending money on miniatures, they were also spending money on themselves by enriching their leisure time and fulfilling their dreams.

> Money spent on the hobby can easily be thought about in relation to the time spent. That I pay e.g. 15 EUR for a miniature and spend approx. 8–20 hours painting it is quite a good relative price (especially if I spend that number of hours several times over on the game the aforementioned mini is part of). I guess 15 EUR is nowadays the norm for a movie ticket...or a couple of pints of beer. (ID 89)

> I spend too much [money] on the hobby, but I'm not out of control. [...] As I've grown older and there's more money, buying has turned more towards shopping for quality than for price. I consider money spent on miniaturing as money spent on mental health and resting your mind and that's valuable :). (ID 72)

Collecting miniatures was sometimes as much about the past than it was about the current moment, and some respondents mentioned buying miniatures out of nostalgia. Here, miniatures took on new meanings: they were links to the past in terms of both the respondents' youth and the pastime itself and reminders of a bygone era (see Williams & Tobin, 2021).

> Furthermore, many of the new minis looked pretty miserable to me; boring and unimaginative. Yes, I'm pretty sure time has gilded and polished my memories and my hunger for nostalgia is not sated by the newest of new things. (ID 45)

> Our gaming group collects so-called oldhammer/middlehammer miniatures created during the 80s–90s. This is in part due to our fondness for the aesthetics of miniatures from that era and in part because it's resistance to contemporary cookie cutter moulded heroes. (ID 82)

Alongside the many positives associated with miniaturing, many respondents identified factors likely contributing to the pile of shame phenomenon. Many recognized impulsive buying behaviours and mentioned situations in which they had overspent on miniatures and supplies. Some respondents explicitly mentioned being aware that they had more miniatures than they needed or had time to paint. There were also mentions of respondents giving themself permission to make new purchases only under given conditions.

> When I restarted miniaturing about ten years ago, I noticed that I was spending quite a lot of money. I started to keep track of my miniaturing hobby costs. When I started to have a lot of minis and less time due to kids, I felt that buying new minis didn't make sense anymore. I started selling off the minis I had collected. From that time I've balanced buying and selling, aiming to rein in my yearly miniaturing budget and keep it comparable to e.g. my physical exercise hobbies. I try to be very rational when it comes to the money (and time!) I spend on minis, but there are a lot of temptations. At times I notice getting new nice stuff exceeding my budget. That's followed by a "remorse period", during which I rationalize my collection, sell off surplus and motivate myself to finish minis. (ID 27)

> For some odd reason there always seem to be more [miniatures] than what you can finish painting. In recent times I've tried to curb 'extra' purchases and rather get something finished instead. It's just the number of projects is quite something already to begin with. My unpainted material consists of maybe a few hundred miniatures (and I believe I represent the more moderate school of hobbyists) but being a hoard...uhmm, collector by nature, I'm not about to give up anything anytime soon. (ID 89)

> I buy almost everything second-hand because the hobby is so very expensive. I currently need to limit my spending a little. Once these miniatures and supplies have been used and the projects are done, I give myself permission to purchase more. (ID 56)

> Usually the things I acquire have been considered so I don't regret them. I'm currently much more deliberate when it comes to my figure purchases. I'm better aware that it's something that I have a passion for, but in the end not enough time or energy. (ID 52)

Some responses revealed our participants' awareness of the fundamentally limited nature of time. The difference between mentions of time between this theme and the *practical sustainability* theme was perspective and scope. Whereas responses in the previous theme discussed the availability of time in everyday life, these responses mentioned time on a more existential level, as a finite resource in the life of a human being.

> I've also considered partially selling my collection, because realistically I'm not going to paint many of them anymore, being in my forties my shaking hands and degrading eyesight give me enough trouble already. (ID 13)

> Today I have almost 1900 figures fit for gaming as well as 500 vehicles in 20mm scale, mostly for the Second World War and subsequent conflicts. In addition I have more unpainted and unbuilt [miniatures and kits] than I'll ever be able to paint during the rest of my life. (ID 69)

Providing an interesting contrast to the practical considerations of the first theme, the examples in this theme highlight the less tangible aspects of collecting miniatures. Here, miniatures are not only physical objects. Instead, they represent investing in oneself and personal well-being, an escape from everyday worries and stress, a nostalgia trip, and an avenue for creativity and self-expression. This personal dimension can also bring with it self-reflection, prompted for example by buyer's remorse after shopping, and lead even to profound considerations of life's priorities.

5 Discussion

Buying and collecting miniatures is an important aspect of miniaturing. Despite some respondents reporting that they were accumulating, or had accumulated, miniatures in disproportionate amounts in relation to their painting output, our data does not suggest pathological hoarding behaviour (see Nordsletten et al., 2013) or compulsive buying of miniatures and miniaturing supplies but rather a high level of engagement with the pastime. The results suggest a similar dynamic to other activities such as digital gaming (Domahidi & Quandt, 2015) or even work (Schaufeli et al., 2008): high engagement shares elements with problematic behaviour on the surface, but there is a difference between the two also in a collecting context (Belk et al., 1991).

The phenomenon of the pile of shame, or the regret and discomfort over the excess accumulation of miniatures, certainly exists. In our data, negotiating with oneself about how and when to purchase new miniatures and justifying these purchases was common. Respondents made sense of their accumulated collection by talking about a past when they were lax about acquisitions but discussed their present habits of purchasing as result of consideration, sometimes according to complicated rules and regulations they had set for themselves.

The two themes explored above illuminate how both the pile of shame and the miniature collection overall are both tangible and intangible. While miniaturists may take a very practical view towards their collections, miniatures are rarely just matter devoid of meaning. Instead, the collection represents an amalgamation of the material and the immaterial. The practical and the existential dimensions both shed light on the piles of shame.

From the practical point of view, the pile of shame is simply excess consumption—the purchase and ownership of miniatures surplus to a miniaturist's personal needs. They represent wasted money that could have been spent on something else and likely cannot be regained by selling them, they take up space that is limited to begin with, and due to their unimportance, time and energy are unlikely to be spent on painting them or playing with them. This kind of surplus can form because of different reasons: sometimes they are a poorly considered purchase from the start, sometimes changes in interest and preferences or newer and better versions render them obsolete. As there is a limited aftermarket and miniatures cannot be easily recycled, often this surplus just ends up sitting in storage, taking up space. The pile of shame limits the pastime in practical terms, rendering it less sustainable.

From the existential point of view, the pile of shame can be seen as a testament to failures of personal character and represents things that render the pastime personally less sustainable. Whether the pile of shame reminds the miniaturist of the excess of impulsive purchases and passed fads or the disappointment of abandoned and forgotten projects, it is less about the practicality and more about the experience: a group of ten untouched miniatures can be a source of guilt despite not taking up much space, not requiring considerable time to paint, and not being particularly expensive. In contrast to the space physically taken up by miniatures, the pile of shame represents miniatures occupying mental space in a negative manner. The playful promise and invitation to fantasy an unpainted miniature offers can turn sour when lack of time, the size of the task ahead, or waned interest crushes the joy out of the anticipation. Miniatures that are not part of imagined future projects start to weight on the miniaturist.

While these observations may sound dramatic, perspective needs to be kept in mind. Although words like “guilt”, “shame”, or “need” are used by miniaturists in our data, the responses as well as the authors’ experiences of miniaturing culture suggest that these are typically not very intense experiences. Although a miniaturist may refer to their excess miniatures as a pile of shame and consider it a negative aspect of their pastime, it is almost certain that this usually represents fairly minor frustration rather than profound feelings of shame and personal failure.

5.1 Life Cycles of Miniatures and Miniaturing

During our study we discerned a life cycle for a miniature from the point of view of a miniaturist,[2] as well as miniaturing as an activity, the two often running parallel. The miniature starts out as a non-tangible dream, story, or idea, and typically ends with the model either being sold onwards or put more or less permanently into storage, whether in a glass cabinet as a display piece or packed into a box with countless others, or very rarely permanently disposed of. In between the miniature finds many uses: it is a monetary expense, an outlet for creativity and self-expression, a tool for stress relief, a toy, a gaming token, an anchoring point for the imagination, and an object taking up space—often many of these at the same time (Meriläinen et al., in press; on toy life-cycles, see Heljakka, 2013, p. 307).

According to our data, miniatures are not only purchased new (or printed new) but are often acquired second-hand and rarely thrown away. This means that the life cycle indeed becomes a cycle, as one miniaturist’s dwindled interest becomes the starting point of another’s new project, ends up as a crowning piece of a collection, or is stripped of paint, cut apart and reassembled into something new. Many miniatures have been in this kind of circulation since their original casting in the early 1980s, a testament to their longevity in terms of materiality, commercial, and sentimental value (see Williams & Tobin, 2021). The materials have their limits, of course, and both plastic and metal alloys eventually start to show degradation after several decades. Miniaturists talk about, for example, “lead rot”, lead oxidization affecting older miniatures that leads to visible miniature decomposition (The Toy Soldier Museum, n.d.).

Running parallel to the miniature’s life cycle is the hobby life cycle: for many of our respondents, interest in miniaturing had waxed and waned. Sometimes miniatures had been sold off or they had sat in storage for years with paints drying up, only for interest to be renewed in a new life situation or with a new source of inspiration. To

[2] The life cycle of the miniature as a physical object, starting from its manufacture and ending in its physical disintegration, would be very different.

illustrate through the first author's personal experience: there is something magical about a miniature being cast in Nottingham, UK in the early 1980s, getting bought at a small-town book store in Finland, finding years of use unpainted in role-playing games and imaginative play, sitting at the bottom of a shoebox for two decades and eventually ending up painted in a display cabinet in 2021, imbued with not only childhood nostalgia, but also contemporary value as an aesthetic object, a collectible, and a gaming piece. The metal miniature as a physical object has remained the same throughout its life cycle, while the author has grown from a newborn baby into an adult and a scholar writing about miniatures, and the world has fundamentally changed around both. On the shelf, miniatures cast from lead alloys during the Cold War march side by side with modern resin figures, 3D printed at home from digital files downloaded off the internet.

As many of our informants have done miniaturing for a long time, decades in most cases, they have a personal relationship to the history of the miniature pastime, miniature communities, and the miniature wargaming industry. They have seen the recycling of the miniatures, as well as dreams and plans relating to the miniatures, many times. They often have a nostalgic attitude towards the past and see childhood as a time when they might not have had all the toys, but they had time to play (see Williams & Tobin, 2021). However, while nostalgia is a significant element and influence, it is not a determining factor. Temporally miniaturing is both something that has happened in the past, informing their taste and style, and an activity occurring simultaneously in their present and future lives. That said, the accumulation of miniatures can also be an indicator of the growth of leisure options and personal finances *and* the reduction of leisure time: while participants have the means to purchase more miniatures, they lack the time to actually paint them. The monetary and temporal resources of the miniaturist go through cycles as well.

This cyclical nature of miniatures and miniaturing holds great relevance for both collecting miniatures and the pile of shame phenomenon. It is common for miniatures to be stored, sometimes for long periods, only to be used again as a new project emerges or when the miniaturist has more time or energy available. During this period of not being in use, a miniature can come to be viewed as being a part of the pile of shame, but this is not necessarily a permanent condition: the appearance of a new project or other suitable use can immediately reframe the miniature as extremely useful instead of surplus—or if it is curated out of the collection, another miniaturist can incorporate it into their collection and imagination. This tension is an important part of the balancing between collecting and amassing a pile of shame and seems to be at the core of suggestions for using the formulation "pile of potential" instead.

5.2 Limitations of the Study

It is important to note that our data on the miniature life cycle as well as the collecting of miniatures is skewed as our informants are self-selected enthusiasts. Miniatures are important to them and they both search for them and appear to care about where their discarded figurines end up. Due to the lack of data, we do not know how representative these respondents are of the general miniaturing population. For example, it is likely that there are numerous people involved in the pastime that are not as reverent of their miniatures or as fastidious with the recycling.

Recycling and re-selling of miniatures is not as organized and systemic as with, say, the circular economy of books or furniture. Stores specializing in selling second-hand miniatures are extremely rare, although there are social media groups devoted to buying and selling, and there is for example a thriving miniature aftermarket on the online auction site eBay. While previously owned miniatures can occasionally be found at thrift stores, flea markets, or used toy stores, it is possible that miniatures do end up in trash more often in the wider miniaturing population than amongst our informants.

Another important observation relates to finances. Miniaturing can require significant financial investment through the purchase of not only the figurines themselves, but also crafting tools, paints, terrain pieces, and rulesets. This presents a very concrete barrier for participation in the pastime and shapes participation for individuals based on their personal finances. For some miniaturists, accumulating quantities of miniatures far beyond their needs is simply not an option. Collecting miniatures is clearly connected to disposable cash.

A number of informants reported that they had taken breaks from the pastime, or significantly altered their engagement with it due to monetary restrictions and worries over the personal and social financial sustainability of the practice. Simultaneously numerous respondents stated that miniaturing is not an expensive pastime, at least not the way they practice it. Most of our respondents were in their 30s or 40s and many of them mentioned working full-time. Several explicitly mentioned having well-paid jobs. This is likely a bias in our data: people who cannot afford to participate in miniaturing are obviously unlikely to do so.

It needs to be underlined that we have chosen to focus on the miniaturists' personal experience, approaching sustainability from the point of view of the individual miniaturist and their hobby practices. Environmentalism or ecological concerns featured seldom in the data, and amassing miniatures was seen more as a practical challenge ("how to store?") or an existential problem ("will I die before all of these are painted?"), than a question of environmental sustainability. The environmental sustainability of a pastime where a single miniaturist can amass collections of thousands of metal and plastic miniatures and the assorted tools and materials is an important topic for future research.

5.3 *From a Pile of Shame to a Curated Collection*

In this chapter and in miniature communities, the accumulation of a superfluous amount of miniatures is discussed as the pile of shame. An unpainted, unconstructed miniature is a promise of many kinds of enjoyment and material for play and self-expression, but it also takes up room both physically at home and mentally as a task to be finished, as excess and waste. However, the pile of shame is also a pile of pride and potential, promising fantasy, relaxation, and play. One day the unpainted miniatures are a joy, another day they feel like an impossible chore. Then it is not the miniaturist who owns the toys, but the toys that own the miniaturist.

Our respondents' experiences suggest that the way to temper the shame of the pile is to curate one's collection. Selecting which miniatures to keep and what to sell off or gift both addresses the practical challenge of managing a large physical collection and helps maintain a selection of miniatures centred on one's own interests. Indeed, curating is what turns an assortment of things into a collection. We can see from our data that our informants place a value on their miniatures that is separate from their value as commodities, as objects for play, as craft material, or even as collectibles. As part of a collection they have an intrinsic value as parts of a larger whole. The miniatures have *culture-value* (see Johnson & Luo, 2019) which goes beyond their use as objects to be gamed with, painted, or displayed (cf. Williams & Tobin, 2021).

The goal of the collector is not to simply amass as many miniatures as possible, but to make conscious, informed choices about new purchases, but also selling off both painted ("used") and unpainted and unconstructed figurines ("unused"). Arguably curating is always a part of collecting, which is has been characterized as the "process of actively, selectively, and passionately acquiring and possessing things removed from ordinary use and perceived as a part of a set of non-identical objects or experiences" (Belk, 1995, p. 67) and as an "active and discerning process" (Geraghty, 2014, p. 14; see also Heljakka, 2017; McIntosh & Schmeichel, 2004).

The curation follows self-created rules—which can be negotiated and bent but still stand—which further civilize, organize, and give meaning to the collection. The collection is imagined by the collector, and when it no longer makes sense, when there are too many miniatures that do not fit current and future plans, re-imagining and pruning are needed. Indeed, during the past years, amateur curation has gained interest in various fields of object interaction from popular music heritage studies (Withers, 2018) to fashion curation (Petrov, 2019). A curator offers an interpretation of how objects relate to one another. Essentially, objects tell a story (Wolff & Mulholland, 2013). For many informants crafting a story with the figurines that is influenced by play and game play is very concrete—and central to the pastime. However, storytelling also happens metaphorically. Collections are in a constant state of becoming; this defines the processual nature of the collecting pastime.

A curated collection, imbued with meaning by the curator, is a reflection of the collector. For some informants, the mass of unpainted figurines acquires transcendental meaning when they state that they know that they will never paint them all before they die. Even in their passing, the miniatures remain, the collection a reflection of its collector. In game cultures some collectors have made plans to turn their collections over to be archived after passing away, this meaningful resignification of game-related objects and the process of donation transforming the relationships with those objects (deWinter & Kocurek, 2017). The plans miniaturists have for their collections after their own death are not addressed in our data, but it is likely that after years of amassing a collection, miniature enthusiasts do not want the life cycle of their miniatures to end when theirs does, and such plans present a fascinating topic for further research.

6 Conclusions

The relationship between the miniaturist and their collection is often a very meaningful one. The pastime of miniaturing is something that provides an opportunity for self-expression and play, for solitude and reflection, and for socially shared and recognized creativity and meaning-making. The key material building blocks of the pastime are commercial objects created for consumption, encased in both shifting trends and nostalgia, and often connected to proprietary intellectual property and its storyworlds. The miniaturist also incorporates materials that would otherwise be considered trash, remixes products from different manufacturers and transmedia worlds, prints, and even designs their own miniatures, and assembles new and personal wholes. The miniaturist's collection is composed of painted and unpainted miniatures, armies, and dioramas, but also of imagined but not yet completed projects and miniatures that are still waiting to be incorporated into an actual or imaginative project. While miniaturing draws on the past and nostalgia as well as completed projects and materials, it is future oriented and moving towards a personal goal and vision. Most importantly, the collection makes and needs to make, sense to the collector.

At times the miniaturist may amass more miniatures than they have plans for, the titular pile of shame, and this produces anxiety and unease. The miniature collection which is supposed to be a positive force in their lives becomes a burden and threatens to become unsustainable. While owning miniatures that are not yet part of a practical project and owning miniatures that are part of an imaginative one is meaningful, having too many miniatures without a meaning is taxing. The miniaturist gains mastery over their collection by pruning and curating it, by rationalizing, planning, and constructing rules. In a word the miniaturist re-imagines the collection and lets go of the miniatures that no longer fit this new vision of their collection and its future. The collection continues to be imagined by the collector until the collector is no more; then only the collection remains.

From the point of view of global ecology and environmental sustainability, it is easy to question a transnational pastime industry centred around plastic miniatures. For the miniaturist, however, their playful leisure pastime is sustainable when it functions as a source of joy. Strictly speaking toys, especially adult toys, do not make sense in the frameworks of efficiency and functionality. Miniatures are, among other things, toys, and toys are frivolous, useless, extra, in a word: waste. Yet at the same time toys are at the heart of humanity. Creating toys marks us apart from other animals, and the connected meaning-making is meaningful. Struggling to keep this pastime as a positive force for the miniaturist in the pressures of limited resources of time, space, and money, social obligations, fads and fashions, capitalistic incentives to overconsumption, and personal limitations is challenging. Even so, when successful, the pile of shame transforms through the collection's re-imagining into a pile of potential.

References

Apperley, T. (2010). *Gaming rhythms: Play and counterplay from the situated to the global*. Institute of Network Cultures.

Arendt, H. (2007). Introduction. In Benjamin, W. *Illuminations. Essays and reflections* (H. Zohn, trans., H. Arendt, ed., pp. 1–55). Schocken Books. (Original work published 1968).

Belk, R. W. (1995). *Collecting in a consumer society*. Routledge.

Belk, R. W., Wallendorf, M., Sherry, J. F., & Holbrook, M. B. (1991). Collecting in a consumer culture. In R. Belk (Ed.), *SV—Highways and buyways: Naturalistic research from the consumer behavior odyssey* (pp. 178–215). Association for Consumer Research.

Benjamin, W. (2007). Unpacking my library. In W. Benjamin, *Illuminations. Essays and reflections* (H. Zohn, trans., H. Arendt, ed., pp. 59–67). Schocken Books. (Original work published 1931).

Braun, V., & Clarke, V. (2006). Using thematic analysis in psychology. *Qualitative Research in Psychology, 3*(2), 77–101.

Braun, V., & Clarke, V. (2012). Thematic analysis. Research designs: Quantitative, qualitative, neuropsychological, and biologicalIn H. Cooper, P. M. Camic, D. L. Long, A. T. Panter, D. Rindskopf, & K. J. Sher (Eds.), *APA handbook of research methods in psychology* (Vol. 2, pp. 57–71). American Psychological Association.

Burghardt, G. M. (2005). *The genesis of animal play: Testing the limits*. The MIT Press.

Carter, M., Gibbs, M., & Harrop, M. (2014). Drafting an army: The playful pastime of Warhammer 40,000. *Games and Culture, 9*(2), 122–147.

Carter, M., Harrop, M., & Gibbs, M. (2014). The roll of the dice in Warhammer 40,000. In F. Mäyrä, K. Heljakka, & A. Seisto (Eds.), *ToDIGRA: Physical and digital in games and play* (pp. 1–28). ETC Press.

Cova, B., Pace, S., & Park, D. J. (2007). Global brand communities across borders: The Warhammer case. *International Marketing Review, 24*(3), 313–329.

Coward-Gibbs, M. (2021). Why don't we play Pandemic? Analog gaming communities in lockdown. *Leisure Sciences, 43*(1–2), 78–84.

davekay. (2020, June 20). Shame, or possibility? [Web log post] Retrieved from https://scentofagamer.wordpress.com/2020/06/20/shame-or-possibility/

deWinter, J., & Kocurek, C. A. (2017). Repacking my library. In M. Swalwell, A. Ndalianis, & H. Stuckey (Eds.), *Fans and videogames: Histories, fandom, archives* (pp. 165–179). Taylor & Francis.

Dhar, U., Liu, H., & Boyatzis, R. E. (2021). Towards personal sustainability: Renewal as an antidote to stress. *Sustainability, 13*(17), 9945.

Domahidi, E., & Quandt, T. (2015). "And all of a sudden my life was gone…": A biographical analysis of highly engaged adult gamers. *New Media & Society, 17*(7), 1154–1169.

Geraghty, L. (2014). *Cult collectors: Nostalgia, fandom and collecting popular culture*. Routledge.

Harrop, M., Gibbs, M., & Carter, M. (2013). Everyone's a winner at Warhammer 40K (or, at least not a loser). In *Proceedings of DiGRA 2013 Conference: DeFragging Game Studies.*

Heljakka, K. (2013). *Principles of adult play(fulness) in contemporary toy cultures: From wow to flow to glow*. Helsinki: Aalto ARTS Books. Aalto University publication series DOCTORAL DISSERTATIONS 72/2013.

Heljakka, K. (2017). Toy fandom, adulthood, and the ludic age. In J. Gray, C. Sandvoss, & C. L. Harrington (Eds.), *Fandom: Identities and communities in a mediated world* (pp. 91–105). New York University Press.

Johnson, M. R., & Luo, Y. (2019). Gaming-value and culture-value: Understanding how players account for video game purchases. *Convergence, 25*(5–6), 868–883.

Kankainen, V. 2016. The interplay of two worlds in Blood Bowl: Implications for hybrid board game design. In *Proceedings of the 13th International Conference on Advances in Computer Entertainment Technology*, Article No. 8. ACM.

Körner, R., & Schütz, A. (2021). It is not all for the same reason! Predicting motives in miniature wargaming on the basis of personality traits. *Personality and Individual Differences, 173*, 110639.

Langer, B. (1989). Commoditoys: Marketing childhood. *Arena, 87*, 29–37.
Lewin, C. G. (2012). *War Games and their History*. Fonthill.
Lundy, J. (2021). *Toying with identity: Adult toy collectors, material fandom, and generational media audiences* [Unpublished doctoral dissertation]. Drexel University.
McIntosh, W. D., & Schmeichel, B. (2004). Collectors and collecting: A social psychological perspective. *Leisure Sciences, 26*(1), 85–97.
Meriläinen, M., Heljakka, K., & Stenros, J. (in press). Lead fantasies: The making, meaning, and materiality of miniatures. In C. Germaine & P. Wake (Eds.), *Material game studies*. Bloomsbury.
Meriläinen, M., Stenros, J., & Heljakka, K. (2020). More than wargaming: Exploring the miniaturing pastime. *Simulation and Gaming*. OnlineFirst.
Nordsletten, A., Reichenberg, A., Hatch, S., De la Cruz, L., Pertusa, A., Hotopf, M., & Mataix-Cols, D. (2013). Epidemiology of hoarding disorder. *British Journal of Psychiatry, 203*(6), 445–452. https://doi.org/10.1192/bjp.bp.113.130195
Peterson, J. (2012). *Playing at the world: A history of simulating wars, people and fantastic adventures from chess to role-playing games*. Unreason Press.
Petrov, J. (2019). *Fashion, history, museums: Inventing the display of dress*. Bloomsbury Academic.
Schaufeli, W. B., Taris, T. W., & van Rhenen, W. (2008). Workaholism, burnout, and work engagement: Three of a kind or three different kinds of employee well-being? *Applied Psychology, 57*(2), 173–203.
Shell, E. R. (2009). *Cheap: The high cost of discount culture*. The Penguin Press.
Singleton, B. E. (2021). "We offer Nuffle a sausage sacrifice on game day". Blood Bowl players' world-building rituals through the lens of theory of sociocultural viability. *Journal of Contemporary Ethnography, 50*(2), 176–201.
Sutton-Smith, B. (2017). *Play for life: Play theory and play as emotional survival*. The Strong.
The Toy Soldier Museum. (n.d.). *Caring for your figures*. Retrieved October 20, 2021, from http://www.the-toy-soldier.com/index.cfm?siteid=259&itemcategory=31947&priorId=31841&pid=31841
Williams, I., & Tobin, S. (2021). The practice of Oldhammer: Re-membering a past through craft and play. *Games and Culture*. OnlineFirst.
Withers, D. M. (2018). DIY institutions and amateur heritage making. In S. Baker, C. Strong, L. Istvandity, & Z. Cantillon (Eds.), *The Routledge companion to popular music history and heritage* (pp. 294–302). Routledge.
WizKids. (2021, January 21). *WizKids, critical role partnership brings Exandria to life*. Retrieved from https://wizkids.com/2021/01/21/wizkids-critical-role-partnership-brings-exandria-to-life/
Wolff, A., & Mulholland, P. (2013, May 1–3). Curation, curation, curation. *Narrative and Hypertext Workshop (NHT'13)*.
Wudugast. (2020, June 20). A right pile of potential [Web log post]. Retrieved from https://convertordie.wordpress.com/2020/06/20/a-right-pile-of-potential/

Wooden Toys Produced from Wood Waste from Urban Afforestation: Acceptance and Implementation Strategies

Luiz Fernando Pereira Bispo, Adriana Maria Nolasco, Gabriela Fontes Mayrinck Cupertino, Fabíola Martins Delatorre, Allana Katiussya Silva Pereira, Elias Costa de Souza, Álison Moreira da Silva, Debora Kilngenberg, José Otávio Brito, and Ananias Francisco Dias Júnior

Abstract In recent years, the use of plastic in the production of toys has increased. Although it is a cheap and long-lasting material, and used as a base for the production of various products worldwide, it is not a sustainable and renewable material and has a long time for decomposition. The unrestrained use of this material can cause several impacts to the environment. Wooden toys have a hereditary and ancient history, as they were part of the culture and pedagogical formation of several generations of children around the world, and the artificialization of toys led to a detachment from this concept. The wastes from tree management of urban forests are commonly used in organic compounds or as firewood, but have great potential for recovery in small wooden objects, such as toys. This is an alternative that adds more value to wood waste and can encourage new ventures, job creation, and income. In this context, it is important to understand the real potential that these objects have on the market, whether they will be well accepted and also what are the most adequate requirements in relation to the expectations of the consuming public. The investigation on the acceptance patterns of children for toys made from wood is necessary for a greater insertion of this sustainable raw material in the market. It was possible to conclude that, based on some improvements regarding the design, choice of species, and the toy manufacturing process, waste from urban forests can be used in the production of these small wooden objects. In this study, we noticed that children of the studied age group are more attracted to colorful toys than natural ones, but that there was good

L. F. P. Bispo · A. M. Nolasco · E. C. de Souza (✉) · Á. M. da Silva · D. Kilngenberg · J. O. Brito
"Luiz de Queiroz" College of Agriculture, University of São Paulo, Piracicaba-SP, Brazil

G. F. M. Cupertino · F. M. Delatorre · A. F. Dias Júnior
Department of Forestry and Wood Sciences, Federal University of Espírito Santo—UFES, Jerônimo Monteiro-ES, Brazil

A. K. S. Pereira
State University of Southwestern Bahia—UESB, Candeias-BA, Brazil

S. S. Muthu (eds.), *Toys and Sustainability*, Environmental Footprints and Eco-design of Products and Processes,
https://doi.org/10.1007/978-981-16-9673-2_5

acceptance of toys manufactured with the wood wastes. The parents and educators of children from 2 to 7 years old also accepted the toys well, but with certain reservations regarding the design and choice of species, which should be suitable for each purpose.

Keywords Sustainable development · Urban wood waste · Toys acceptance · Child education · Child development · Teaching tools

1 Introduction

Playing is considered an abundant and rewarding activity for a child's development. Playing is an activity valued in all cultures. Several studies address the importance of playing for child development. Attractive and natural, this activity promotes cognitive, physical, and emotional well-being, favoring children's development and learning (Maynard & Waters, 2007; Walker & McKay, 2021). Creativity, problem solving, and cooperation are some of the childhood lessons provided during playing. It is easy to see that in an environment with children, playing is an excellent form of communication practice, favoring their social development, and the main instrument of children's plays is the toy. Ball, doll, cart, shuttlecock, games, among other toys, play a fundamental role in the development of language, cognition, and motor coordination.

Within this universe of games, the innovations that ensure greater attractiveness are countless, and it has worked. The toy sector was responsible for the turnover of about US$95 billion in 2020 (TIE, 2021). However, the search for objects that arouse the desire of children does not always have a sustainable character. That is because 90% of toys are produced from plastic materials (Nur-A-Tomal et al., 2020). It is almost unimaginable to think of a current toy that does not have plastic as its raw material. And which is the main element affected by this problem? The environment. With the high volume of plastic production and its relatively short useful life, when evaluating its durability and persistence in nature, plastic has caused a huge negative impact on our planet (Hossain et al., 2021). And it is within this perspective that alternatives aimed at the generation of toys from sustainable materials have been identified as a potential possibility to mitigate the impacts caused by the toy sector.

With sustainable, durable, and ecological aspects, wooden toys are an excellent alternative to mitigate the impacts caused by the toy industry, in addition to enabling children to assimilate the values of environmental protection. In this context, urban afforestation waste is being considered a potential opportunity for the generation of ecological toys (Bispo et al., 2021). Within this context, environmental education should be highlighted as a practice that promotes transformation, in which individuals are encouraged to be agents in promoting sustainable development. In addition to enabling the valorization of these waste materials, which are most often sent to landfills, they ensure the promotion of a novelty in the market. However, to assess the potential of this alternative for the market, acceptance aspects must be taken into account. In agreement with this, this work aims to evaluate the aspects that can contribute to the use of urban afforestation waste as an alternative for the generation of wooden toys. Initially, we will address the main aspects related to the use of toys in children's development. We deal with the toy industry and the main problems related to the use of plastic to produce toys. Finally, we evaluated the acceptance of wooden toys by parents, children, and teachers involved in the children's initial educational training.

2 Toys and Child Development

From ancient philosophy to modern scientific research, playing has been seen as an integral component of child development (Kamal et al., 2017). In general, society, whatever the time, has always valued the practice of playing, which is considered important for children's learning (Kärtner & von Suchodoletz, 2021; Roopnarine et al., 2018). Playing is beneficial for neurological and physical development, making great contributions to the development of a child's motor, social, and emotional skills (Goldstein, J., & Goldstein, 2012; Hall et al., 2021; Weisberg et al., 2013). Studies claim that children build knowledge as they explore their world, with playing being the main support for achieving various milestones in cognitive development (Fehr et al., 2020; Hall et al., 2021). Considered essential in the first years of life, the involvement of children in plays favors the development of imagination, creativity, and effective problem-solving skills. In addition, playing is an important form of communication, and it is through this act that children can reproduce their daily lives. Considered a creative human activity, playing makes imagination, fantasy and reality interact, producing new ways of building social relationships with other subjects. Every child can learn by playing. This reality drives the inclusion of playful learning activities, enabling learning without difficulty, promoting the understanding of educational content (Akdeniz & Özdinç, 2021). This context fits the phrase of Paulo Freire: "To educate yourself is to impregnate with meaning every moment of life, every everyday act."

Playing is recognized worldwide as an essential part of learning and growth, and as toys are an essential tool in the practice of playing, they are invaluable for a child's development (Bedford, 2021). Toys are an important instrument to stimulate imagination and motivation, favoring cognitive and motor skills, in addition to enabling the transmission of teachings of cooperation, communication, and sharing (TIE, 2021). Defined as an object for children's play (made or purchased), it has a great impact on the development of children's domains (including cognitive, language, socio-emotional, and physical) (Goldstein, J., & Goldstein, 2012; Healey & Mendelsohn, 2019). Physical toys are categorized as symbolic, such as dolls and cars, as adaptive, such as puzzles, and as related to language, such as cards and board games (Choi et al., 2018). Although the concept of playing has not changed over time, what constitutes a toy these days is different from what it was centuries ago (Roopnarine et al., 2018). This difference is based on the proliferation of light toys, with sensory stimulation and digital platforms (Richards et al., 2020).

In recent years, several elements of traditional toys have been adapted to technology-encompassing versions, such as laptops, tablets, phones, other mobile devices, and standalone electronic game devices, which end up replacing human interaction (Levin & Rosenquest, 2001). The evolution of technology is evident every day, and it arises along with these new ways of playing, whether through new games, online platforms, and devices. Most controversially, we are in an era when psychologists and developmental scientists are proving the body's role in human learning, and children's toys are becoming increasingly two-dimensional (Hastings et al., 2009). This whole issue ends up interfering in the way the generation plays and this needs to be problematized in schools and society in general. We highlight a concern here: What are the implications of this for the child's development? Are traditional toys losing their place because they are not so attractive and competitive?

3 The Toy Industry

Despite the great challenges faced due to the insertion of technology in children's play, the toy industry is still growing. In the year 2020, the global market for the toy industry had a growth of about 2.6% compared to 2019 (TTA, 2020). It is considered a homogeneous industry, with a predominance of small and medium-sized companies, controlled by a small group of large internationalized companies, which carried out the externalization of large-scale production to other countries, mainly in Asia. In general, the greatest concentration of toy production is located in China, which accounts for about 70% of world production, followed by France, Germany, Italy, and Spain (ABDI, 2011), characterized by the high concentration in the export scenario, reaching approximately US$ 33 billion in 2020, 17% higher than 2019, with the United States and the European Union being the main consumers (TTA, 2020). The toy industry is huge.

Even with the insertion of technology, one feature is of paramount importance for the success of the toy industry: innovation. Based on this differential, the sector manages to introduce new products in the commerce, satisfying the needs and desires of consumers. However, often in search of better acceptance, the toy industry ends up adhering to the production of unsustainable materials, causing a huge negative impact on the environment. Studies claim that toys made with plastic materials represent 90% of total toy sales in the world (Goldstein, J., & Goldstein, 2012; Nur-A-Tomal et al., 2020). Yes, toys are designed to arouse joy, but when children's interests change, they turn into garbage. And thinking about sustainability, the situation is worrying.

4 Environmental Problems of Plastics

Light, durable, and low cost are the main characteristics that make plastic materials increasingly ingrained in the daily life of society. A world without plastics seems unimaginable these days. Yes, plastic materials have infiltrated almost every aspect of human life, and the production numbers are staggering. A survey carried out by Plastics Europe elucidates that in 1950 the global volume of plastics production was 2 million metric tons, while in 2016 the production was 317 million metric tons, a volume almost 160 times higher (Brooks et al., 2018; Plastics Europe, 2017). This intense production ends up causing serious problems to the environment, contributing to the generation of about 6 billion metric tons of plastic waste since 1950 (Geyer et al., 2017). The situation is worrying. Of all this volume generated, around 85% is disposed of in landfills or into the environment, representing serious threats to society and the environment (Liu et al., 2021). Studies warn that about 8 million metric tons of plastic waste enter the oceans annually and estimates indicate that if the situation is not reversed, basically 99% of seabirds will have ingested plastic waste by the year 2050 (Jambeck et al., 2015; Thompson, 2004; UN, 2018). And the situation does not stop there, the production of plastic materials also favors the emission of greenhouse gases. This is because 90% of plastic products are based on virgin oil, which corresponds to 20% of oil production destined for the generation of this material (Ellen Macarthur Foundation, 2016; Liu et al., 2021; Maynard & Waters, 2007; Walker & McKay, 2021).

The generation of plastics, as evidenced by the islands that currently float in the Pacific Ocean, is plentiful. And in the toy industry, it is no different. The reasons why most toys are made from plastic materials are related to their durability, strength, and the ability to be washable, which are advantageous characteristics for the generation of a children's product (Krosofsky, 2021). About 40 tons of plastics for every $1 million in revenue are used in the toy industry. It is the industry that most uses plastic in the world. Polyethylene, polystyrene, polypropylene, acrylonitrile butadiene styrene, polyvinyl chloride, among others, are the polymers that generate the main instrument of children's play (Fig. 1). And what about the environmental impacts? It can be said that they are much larger than a child's imagination can conceive.

It is the benefits of plastic in the production of toys that make it a major and worrying environmental liability (Krosofsky, 2021). Of all toys sold, about 80% end up in landfills, incinerators, or the ocean (Ellen MacArthur Foundation, 2020). Considered short-lived materials, the easiest destination is the disposal in the environment. And what are the negative impacts of this? Several. When discarded in the environment, plastic waste is responsible for the deterioration of the natural environment, which can cause soil and water contamination, as its composition is full of chemical additives that, when exposed to the weather and environment, are leached. (Chamas et al., 2020; Kremer et al., 2021; Okunola et al., 2019). Even though plastic toys are an excellent alternative aimed at the child's development through playing, their production and disposal are considered problematic. Because of this reality, it is

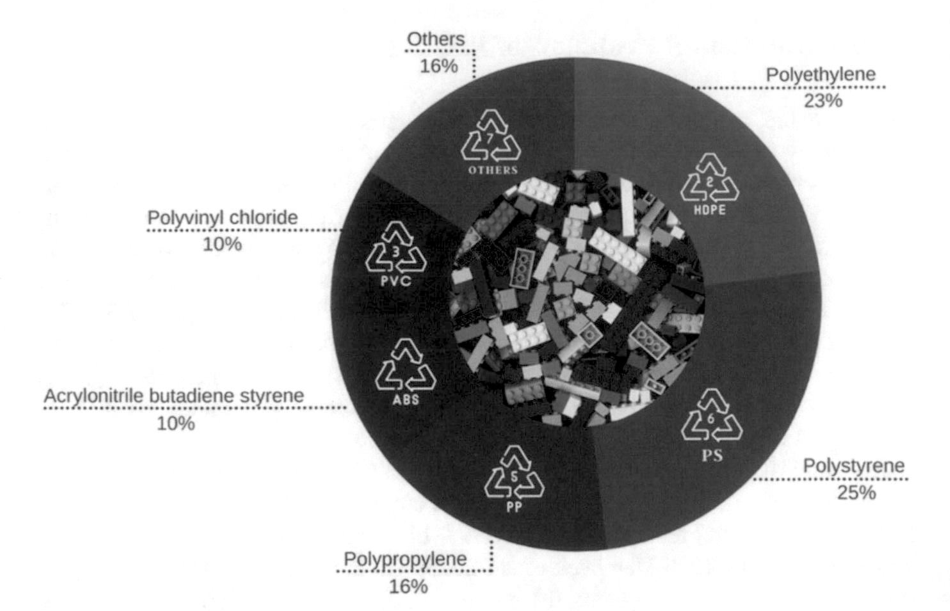

Fig. 1 Polymers used in the toy sector (*Source* BPF, 2010 adapted)

necessary to invest in sustainable materials, aiming to reduce the amount of residual plastic, contributing to the environment.

5 Toys and Sustainability

Replacing the materials that cause environmental problems with those deemed ecological is a challenge in the toy industry around the world. Some companies that sell this material are starting to create policies aimed at circular economy, aiming at the use of 100% recycled, recyclable, or bio-based plastic materials. However, the situation is still far from becoming reality. With a change in production, the price of products can go up, leading to a lack of interest in buying toys made with ecological plastics, which makes us go back to square one. The best alternative is searching for biodegradable and abundant materials. It is in this perspective that it is worth remembering wooden toys, those which a few years ago made the children happy, in addition to often having a sentimental character.

The use of wooden toys is no longer as ingrained in society as it was years ago. Although some traditional physical toys are timeless, children now spend a lot of their time engaging and learning with digital devices. In the twenty-first century, there was a trend toward the production of diversified and globalized toys and technological devices that made wooden toys increasingly distant from the market and children's plays (Hall et al., 2021; Tiwasing & Sahachaisaeree, 2012). Recent trends in reducing the use of plastic and the principles of sustainable development and ecological products allow wood to return to being one of the focuses and materials of interest, as toys made from wood are very durable, solid, organic, resistant, and light, according to the species chosen for manufacturing, in addition to being made with a biodegradable and renewable material (Bispo et al., 2021; Ebner & Petutschnigg, 2007; Tiwasing & Sahachaisaeree, 2012).

In parallel to this, there is a growing need for environmental awareness regarding the materials used for the development of toys, so the use of raw material wood appears as a viable alternative both economically and environmentally. The development and abstract reflection happen due to the connections that the child creates not with the objects themselves but in the coordination of the actions or operations of the activities, such as with the objects in their hands. The stimulation of play awakens in children reflections, connections, feelings, ideas, reasoning, movements, and senses (Cywa & Wacnik, 2020), and a toy made with wood can effectively stimulate "playing" through the human-nature interaction, bringing the child closer to a natural material, used for hundreds of years in child education and development.

6 Waste from Urban Afforestation as an Alternative for the Generation of Toys

Trees in the urban environment can ease temperatures, reduce the impact of rain on the soil, sequester, and store carbon, in addition to their aesthetic function of providing well-being and improving the health of the population, but despite their benefits, their wastes are underutilized resources (Đerčan et al., 2012; Joshi et al., 2015; Martinez Lopez et al., 2020). Silvicultural treatments generate large volumes of woody wastes (trunks, branches, roots) (Martinez Lopez et al., 2020). In Brazil, there is no precise information on the amount of urban wood waste generated, but partial data from the National Sanitation Information System (SNIS) indicates that approximately 97,703.8 tons of waste from pruning and felling are generated annually (SNIS, 2019). In the United States, around 33 million tons of dry wood are generated per year (Bispo et al., 2021; Nowak et al., 2019).

Urban afforestation wastes are often disposed of in landfills or in areas of irregular disposal, often after burning to reduce the volume, which increases air pollution, especially in developing countries (Fetene et al., 2018; Singh et al., 2014). Urban wood waste serves as a source of clean energy, the production of small objects and the manufacture of furniture, aiming at efficient and effective use of the material and generation of income. Reusing this waste is a recent segment in the world, and even if in short steps, some studies have been conducted to provide support and greater added value to this raw material that is so abundant, cheap, and easy to handle. However, to make this alternative viable, it is necessary to characterize the wood of the species used in urban afforestation, with the development of product tests, labor training, investment in production infrastructure, development of a waste screening plan, and studies that assess the drying and storage of adequate marketing strategy (Kamal et al., 2017). It will bring solutions for companies and city halls that need to deal with the correct disposal of material as an alternative to discarding wood. One of the viable alternatives among small wooden objects is the production of toys, helping to minimize environmental impacts and generate income.

Using urban forest wastes as a raw material for the production of toys is an efficient way to value these materials, precisely contributing to mitigating the environmental impacts arising from inappropriate disposal and potential for the development of new ventures, thus improving the development of the site, increasing the generation of jobs and income (Bispo et al., 2021; Faraca et al., 2019). This research addresses relevant points for businessmen and society, as studies that reported the use of waste from urban afforestation as production of toys are very outdated.

7 Acceptance Aspects

For conducting the opinion survey, a structured interview was used to analyze the consumer's preference for toys made from urban tree waste and their acceptance in general. To have a more comprehensive sampling scenario, the research was carried out in two early childhood education schools in the municipality of Piracicaba (São Paulo, Brazil) and in the park of the "Luiz de Queiroz" College of Agriculture of the University of São Paulo (ESALQ, USP). There are three categories of respondents: parents, children, and teachers. The interviews involved both the opinion of the interviewees and the assessment of the acquisition, visual, and cognitive aspects of toys. In all, 16 parents, 34 children (2–7 years old), and 6 teachers were interviewed. The teachers and parents that participated in the interview signed an Informed Consent Form.

In addition to the size of the market and advertising, the selection of toys by caregivers has the potential to enhance their children's play and learning aspects. (Hassinger-Das et al., 2021). Toys produced from non-renewable sources are taking up more and more space, due to cultural distance, social media, and lack of knowledge of new ways to play. Figure 2 investigates if the children had wooden toys and if they would like to get one.

Figure 2 shows that the vast majority of respondents have never had wooden toys, but that they would like to. This demonstrates that the acceptance of toys from urban afforestation is a large segment, in addition to encouraging caregiver-child interaction. When the questions were directed to the children's parents, 30% reported that they never gave wooden toys to the children. When asked about color preference (Fig. 3), only 10% of the children who were interviewed reported preferring the natural color of wood; in contrast, 90% said they preferred colored toys, indicating that the rudimentary look, despite valuing the natural shapes of wood, does not meet the children's preference.

The preference for colorful toys may be related to the visual stimulus that the colors red, blue, green, yellow, and black can provide to children. It was found in

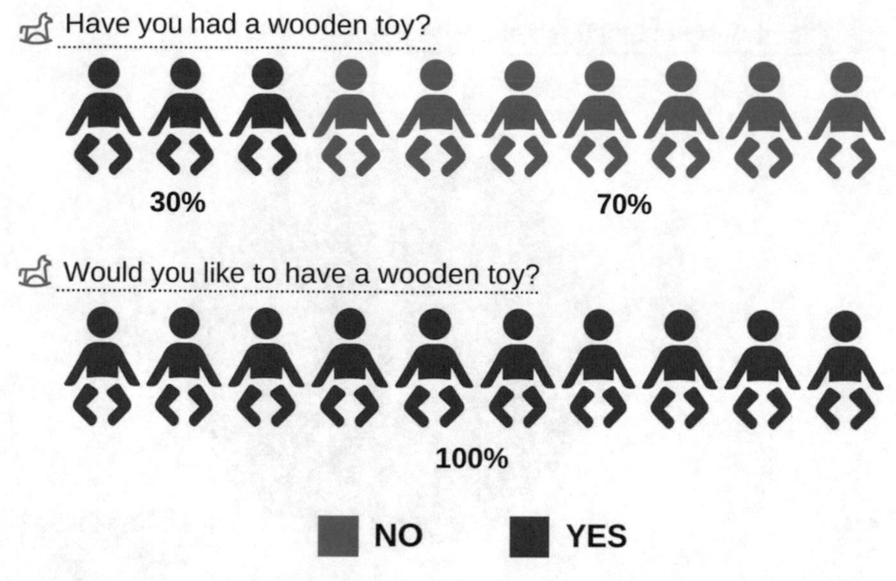

Fig. 2 Research on the purchase of wooden toys (*Source* Authors, 2021)

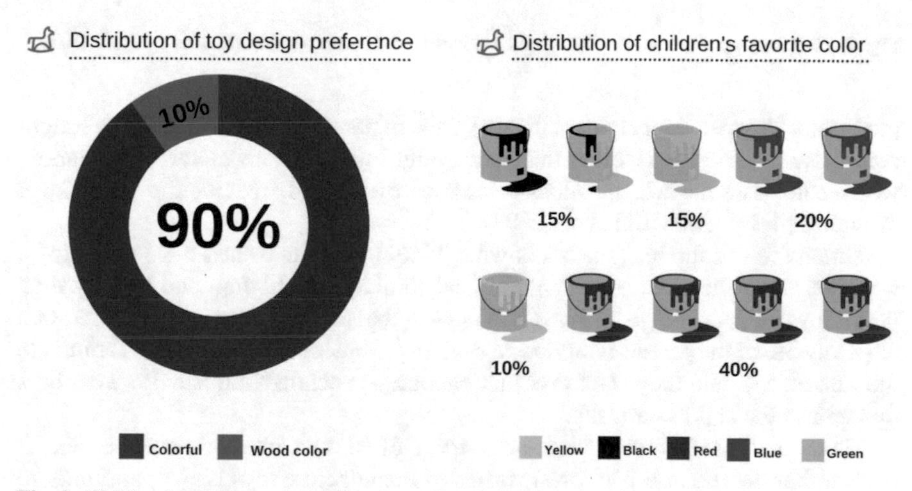

Fig. 3 Children's preference regarding the color of wooden toys (*Source* Authors, 2021)

studies that children and adults, regardless of gender, appreciate the colors red and blue, as they are related to positive emotional aspects (Hassinger-Das et al., 2021; Manches & Plowman, 2021), and they were the favorite colors of most children. Figure 4 shows children's preferences for wooden toys.

When in contact with toys from urban afforestation waste, of the six toys, no vote was given by the children to the village; in contrast, the dolls were preferred by the children with 29% and the train with 26%. As for the cognitive item sought in the

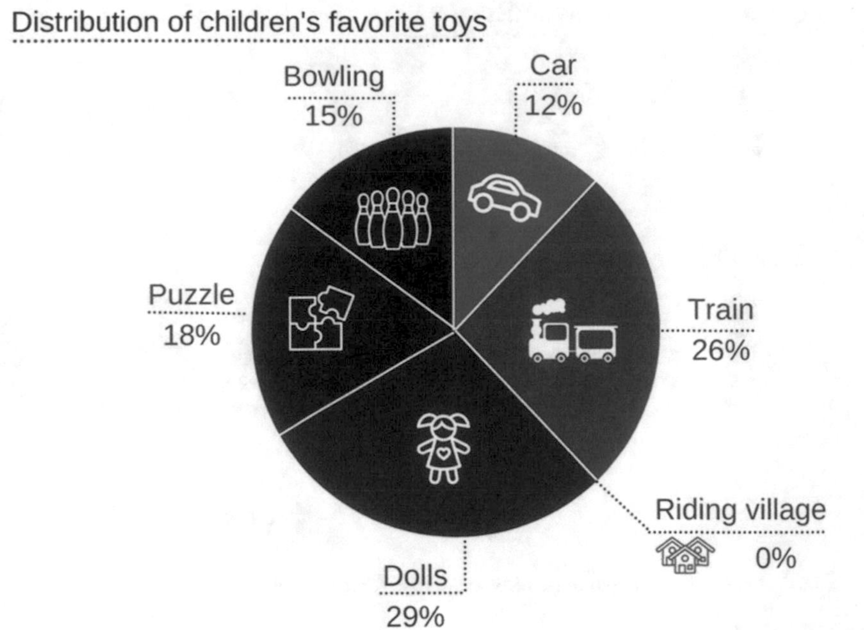

Fig. 4 Children's preference for toys produced from urban afforestation (*Source* Authors, 2021)

purchase of toys for their children (Fig. 5), 29% of parents responded that they sought creativity, while 24% reported that they sought to stimulate motor coordination, which comprises the acts or actions performed by the body that require a high level of brain activity (Pellegrini et al., 2007).

Stimulation of the imagination is what 14% of parents want when buying toys, while 9% want toys that encourage logical thinking in children and 9% say they prefer toys that encourage learning, that is, toys that contain letters, numbers, colors, etc. Only 5% of the parents interviewed said they were looking for toys that stimulate the senses; 5% said they want toys that encourage socialization and 5% want toys that arouse the child's curiosity.

Therefore, based on the acceptance survey of parents, children, and teachers, it is clear that the use of urban forest wastes to manufacture toys is an opportunity to minimize negative environmental impacts and develop new businesses based on this raw material. The valorization strategy occurs through the union of the playful aspect with the environmental one, where in addition to all the characteristics specific to the production of toys, the visual analysis by consumers is also significantly important to validate this reuse.

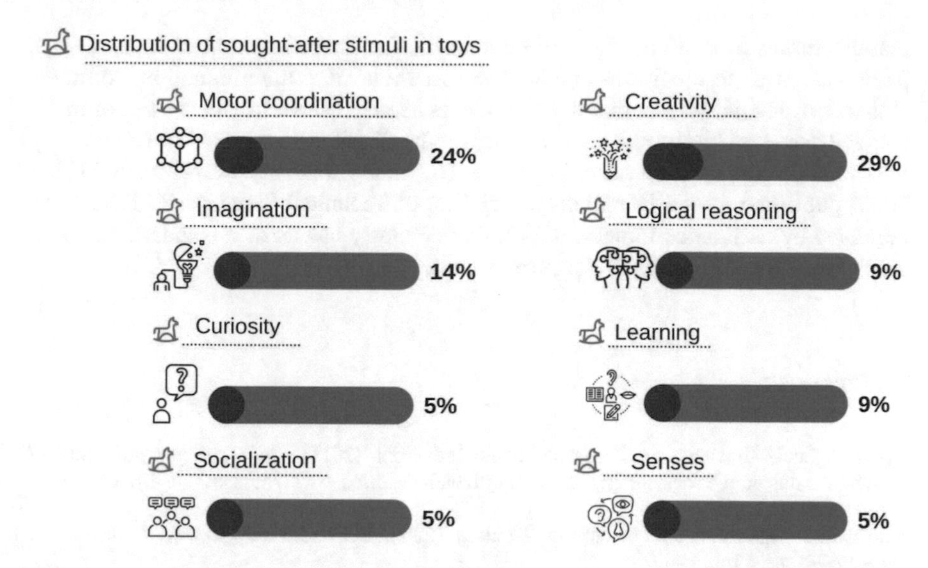

Fig. 5 Parents' purchasing preference for wooden toys in relation to their cognitive function for their children's development (*Source* Authors, 2021)

8 Practical Implications and Future Perspectives

Urban trees bring countless benefits to society, but after removal, this resource is underutilized and often considered a waste to be discarded. Due to the large amount, in most municipalities, these wastes are incinerated or disposed of in landfills. As far as it is known, no technical-scientific study has been carried out analyzing the use of tree pruning wastes to manufacture toys. However, there are companies that produce wooden toys for children, to encourage their development, but their raw material comes from solid wood or reconstituted panels. The expansion of the reuse of these wastes for the production of toys can reduce costs and environmental impacts, such as dependence on fossil sources and use of plastics, and consequently reduce greenhouse gas emissions.

This research study has practical and policy implications as it demonstrates the potential benefits of using urban waste from tree pruning for the production of toys as part of the proper waste management program as well as adequate quality control. In Brazil, these actions are provided for in the National Solid Waste Policy (Law 12305/2010), which encourages recycling and reuse of waste (Brasil, 2009, 2010). The pruning and the removal of trees carried out in most cities are a strategic socio-environmental alternative to serve day care centers and schools, mainly low-income ones, due to the low cost, ease of production and durability of toys, promoting social, intellectual, and motor development of children.

It is important to know the peculiarities of each species, as some chemical components, especially extractives, can vary between species and between the same species depending on the planting location. Therefore, it is of paramount importance before manufacturing to find out if it will be suitable for the production of toys, always paying attention to the norms and laws in force related to the production, commercialization, and adaptation that wood products need to follow in order to be commercialized and used by the end-user. For toys to be distributed and sold in Brazil, they must contain certifications according to the Toy Safety standard (ABNT NBR 11786, 2003) published by the Brazilian Association of Technical Standards (ABNT) and regulated by Ordinance Inmetro n°177. Each country has its own standard, and it is crucial to know it and follow the current laws.

References

ABDI, Agência Brasileira de Desenvolvimento Industrial. (2011). *Toy industry report*. https://www.eco.unicamp.br/neit/images/stories/arquivos/Relatorios_NEIT/Industria-de-Brinquedos-Agosto-de-2011.pdf

ABNT, Associação Brasileira de Normas Técnicas. (2003). *NBR 11786. Segurança do Brinquedo - Especificações* (pp. 1–65).

Akdeniz, M., & Özdinç, F. (2021). Maya: An artificial intelligence based smart toy for pre-school children. *International Journal of Child-Computer Interaction, 29*, 100347. https://doi.org/10.1016/j.ijcci.2021.100347

Bedford, E. (2021). *Receita total do mercado mundial de brinquedos em 2020*. https://www.statista.com/statistics/194395/revenue-of-the-global-toy-market-since-2007/

Bispo, L. F. P., Nolasco, A. M., de Souza, E. C., Klingenberg, D., & Dias Júnior, A. F. (2021). Valorizing urban forestry waste through the manufacture of toys. *Waste Management, 126*, 351–359. https://doi.org/10.1016/j.wasman.2021.03.028

BPF - British Plastics Federation. (2010). Biodegradable polymers in the toy sector. https://www.bpf.co.uk/article/biodegradable-polymers-in-the-toy-sector-364.aspx

Brooks, A. L., Wang, S., & Jambeck, J. R. (2018). The Chinese import ban and its impact on global plastic waste trade. *Science Advances, 4*(6), eaat0131. https://doi.org/10.1126/sciadv.aat0131

Chamas, A., Moon, H., Zheng, J., Qiu, Y., Tabassum, T., Jang, J. H., Abu-Omar, M., Scott, S. L., & Suh, S. (2020). Degradation rates of plastics in the environment. *ACS Sustainable Chemistry & Engineering, 8*(9), 3494–3511. https://doi.org/10.1021/acssuschemeng.9b06635

Choi, J. H., Mendelsohn, A. L., Weisleder, A., Cates, C. B., Canfield, C., Seery, A., Dreyer, B. P., & Tomopoulos, S. (2018). Real-world usage of educational media does not promote parent-child cognitive stimulation activities. *Academic Pediatrics, 18*(2), 172–178. https://doi.org/10.1016/j.acap.2017.04.020

Cywa, K., & Wacnik, A. (2020). First representative xylological data on the exploitation of wood by early medieval woodcrafters in the Polesia region, southwestern Belarus. *Journal of Archaeological Science: Reports, 30*, 102252. https://doi.org/10.1016/j.jasrep.2020.102252

Đerčan, B., Lukić, T., Bubalo-Živković, M., Đurđev, B., Stojsavljević, R., & Pantelić, M. (2012). Possibility of efficient utilization of wood waste as a renewable energy resource in Serbia. *Renewable and Sustainable Energy Reviews, 16*(3), 1516–1527. https://doi.org/10.1016/j.rser.2011.10.017

Ebner, M., & Petutschnigg, A. J. (2007). Potentials of thermally modified beech (Fagus sylvatica) wood for application in toy construction and design. *Materials & Design, 28*(6), 1753–1759. https://doi.org/10.1016/j.matdes.2006.05.015

Ellen Macarthur Foundation. (2016). *The new plastics economy: Catalysing action*. https://ellenmacarthurfoundation.org/the-new-plastics-economy-catalysing-action

Ellen MacArthur Foundation. (2020). *Creating a circular economy for toys*. https://medium.com/circulatenews/creating-a-circular-economy-for-toys-9c11dc6a6676

Faraca, G., Boldrin, A., & Astrup, T. (2019). Resource quality of wood waste: The importance of physical and chemical impurities in wood waste for recycling. *Waste Management, 87*, 135–147. https://doi.org/10.1016/j.wasman.2019.02.005

Fehr, K. K., Boog, K. E., & Leraas, B. C. (2020). Play behaviors: Definition and typology. In *The encyclopedia of child and adolescent development* (pp. 1–10). Wiley. https://doi.org/10.1002/9781119171492.wecad272

Fetene, Y., Addis, T., Beyene, A., & Kloos, H. (2018). Valorisation of solid waste as key opportunity for green city development in the growing urban areas of the developing world. *Journal of Environmental Chemical Engineering, 6*(6), 7144–7151. https://doi.org/10.1016/j.jece.2018.11.023

Geyer, R., Jambeck, J. R., & Law, K. L. (2017). Production, use, and fate of all plastics ever made. *Science Advances, 3*(7), e1700782. https://doi.org/10.1126/sciadv.1700782

Goldstein, J., & Goldstein, J. (2012). *Well-being about the author*. www.kijkwijzer.nl

Hall, L., Paracha, S., Flint, T., MacFarlane, K., Stewart, F., Hagan-Green, G., & Watson, D. (2021). Still looking for new ways to play and learn… Expert perspectives and expectations for interactive toys. *International Journal of Child-Computer Interaction*, 100361. https://doi.org/10.1016/j.ijcci.2021.100361

Hassinger-Das, B., Quinones, A., DiFlorio, C., Schwartz, R., Talla Takoukam, N. C., Salerno, M., & Zosh, J. M. (2021). Looking deeper into the toy box: Understanding caregiver toy selection decisions. *Infant Behavior and Development, 62*, 101529. https://doi.org/10.1016/j.infbeh.2021.101529

Hastings, E. C., Karas, T. L., Winsler, A., Way, E., Madigan, A., & Tyler, S. (2009). Young children's video/computer game use: Relations with school performance and behavior. *Issues in Mental Health Nursing, 30*(10), 638–649. https://doi.org/10.1080/01612840903050414

Healey, A., & Mendelsohn, A. (2019). Selecting appropriate toys for young children in the digital era. *Pediatrics, 143*(1), e20183348. https://doi.org/10.1542/peds.2018-3348

Hossain, M. U., Ng, S. T., Dong, Y., & Amor, B. (2021). Strategies for mitigating plastic wastes management problem: A lifecycle assessment study in Hong Kong. *Waste Management, 131*, 412–422. https://doi.org/10.1016/j.wasman.2021.06.030

Jambeck, J. R., Geyer, R., Wilcox, C., Siegler, T. R., Perryman, M., Andrady, A., Narayan, R., & Law, K. L. (2015). Plastic waste inputs from land into the ocean. *Science, 347*(6223), 768–771. https://doi.org/10.1126/science.1260352

Joshi, O., Grebner, D. L., & Khanal, P. N. (2015). Status of urban wood-waste and their potential use for sustainable bioenergy in Mississippi. *Resources, Conservation and Recycling, 102*, 20–26. https://doi.org/10.1016/j.resconrec.2015.06.010

Kamal, K., Qayyum, R., Mathavan, S., & Zafar, T. (2017). Wood defects classification using laws texture energy measures and supervised learning approach. *Advanced Engineering Informatics, 34*, 125–135. https://doi.org/10.1016/j.aei.2017.09.007

Kärtner, J., & von Suchodoletz, A. (2021). The role of preacademic activities and adult-centeredness in mother-child play in educated urban middle-class families from three cultures. *Infant Behavior and Development, 64*, 101600. https://doi.org/10.1016/j.infbeh.2021.101600

Kremer, I., Tomić, T., Katančić, Z., Erceg, M., Papuga, S., Vuković, J. P., & Schneider, D. R. (2021). Catalytic pyrolysis of mechanically non-recyclable waste plastics mixture: Kinetics and pyrolysis in laboratory-scale reactor. *Journal of Environmental Management, 296*, 113145. https://doi.org/10.1016/j.jenvman.2021.113145

Krosofsky, A. (2021). *Plastic toys have a greater impact on the environment and human health than we thought*. https://www.greenmatters.com/p/environmental-impact-plastic-toys

Levin, D. E., & Rosenquest, B. (2001). The increasing role of electronic toys in the lives of infants and toddlers: Should we be concerned? *Contemporary Issues in Early Childhood, 2*(2), 242–247. https://doi.org/10.2304/ciec.2001.2.2.9

Liu, Z., Liu, W., Walker, T. R., Adams, M., & Zhao, J. (2021). How does the global plastic waste trade contribute to environmental benefits: Implication for reductions of greenhouse gas emissions?

Journal of Environmental Management, 287, 112283. https://doi.org/10.1016/j.jenvman.2021.112283

Manches, A., & Plowman, L. (2021). Smart toys and children's understanding of personal data. *International Journal of Child-Computer Interaction, 30*, 100333. https://doi.org/10.1016/j.ijcci.2021.100333

Martinez Lopez, Y., Paes, J. B., Gustave, D., Gonçalves, F. G., Méndez, F. C., & Theodoro Nantet, A. C. (2020). Production of wood-plastic composites using cedrela odorata sawdust waste and recycled thermoplastics mixture from post-consumer products—A sustainable approach for cleaner production in Cuba. *Journal of Cleaner Production, 244*, 118723. https://doi.org/10.1016/j.jclepro.2019.118723

Maynard, T., & Waters, J. (2007). Learning in the outdoor environment: A missed opportunity? *Early Years, 27*(3), 255–265. https://doi.org/10.1080/09575140701594400

Nowak, D. J., Greenfield, E. J., & Ash, R. M. (2019). Annual biomass loss and potential value of urban tree waste in the United States. *Urban Forestry & Urban Greening, 46*, 126469. https://doi.org/10.1016/j.ufug.2019.126469

Nur-A-Tomal, M. S., Pahlevani, F., & Sahajwalla, V. (2020). Direct transformation of waste children's toys to high quality products using 3D printing: A waste-to-wealth and sustainable approach. *Journal of Cleaner Production, 267*, 122188. https://doi.org/10.1016/j.jclepro.2020.122188

Okunola A. A., Kehinde I. O., Oluwaseun, A., & Olufiropo E, A. (2019). Public and environmental health effects of plastic wastes disposal: A review. *Journal of Toxicology and Risk Assessment, 5*(2). https://doi.org/10.23937/2572-4061.1510021

Pellegrini, A. D., Dupuis, D., & Smith, P. K. (2007). Play in evolution and development. *Developmental Review, 27*(2), 261–276. https://doi.org/10.1016/j.dr.2006.09.001

Plastics Europe. (2017). *Plastics—The factas*. https://www.plasticseurope.org/application/files/5715/1717/4180/Plastics_the_facts_2017_FINAL_for_website_one_page.pdf

Richards, M. N., Putnick, D. L., & Bornstein, M. H. (2020). Toy buying today: Considerations, information seeking, and thoughts about manufacturer suggested age. *Journal of Applied Developmental Psychology, 68*, 101134. https://doi.org/10.1016/j.appdev.2020.101134

Roopnarine, J. L., Yildirim, E. D., & Davidson, K. L. (2018). Mother–child and father–child play in different cultural contexts. In *The Cambridge handbook of play* (pp. 142–164). Cambridge University Press. https://doi.org/10.1017/9781108131384.009

Singh, J., Laurenti, R., Sinha, R., & Frostell, B. (2014). Progress and challenges to the global waste management system. *Waste Management & Research: The Journal for a Sustainable Circular Economy, 32*(9), 800–812. https://doi.org/10.1177/0734242X14537868

SNIS. (2019). Sistema Nacional de Informações sobre Saneamento. (2020). *Diagnóstico do manejo de Resíduos Sólidos Urbanos - 2019*.

Thompson, R. C. (2004). Lost at sea: Where is all the plastic? *Science, 304*(5672), 838–838. https://doi.org/10.1126/science.1094559

TIE, T. I. of E. (2021). *Toys: The tools of play*. https://www.toyindustries.eu/priorities/importance-of-play/

Tiwasing, W., & Sahachaisaeree, N. (2012). Distinctive design perception: A case of toy packaging design determining children and parents' purchasing decision. *Procedia—Social and Behavioral Sciences, 42*, 391–398. https://doi.org/10.1016/j.sbspro.2012.04.203

TTA, T. T. A. (2020). *Global sales data*. https://www.toyassociation.org/ta/research/data/population/toys/research-and-data/data/global-sales-data.aspx.

UN, E. programme. (2018). *Single-use plastics: A roadmap for sustainability*. https://www.unep.org/resources/report/single-use-plastics-roadmap-sustainability

Walker, T. R., & McKay, D. C. (2021). Comment on "five misperceptions surrounding the environmental impacts of single-use plastic." *Environmental Science & Technology, 55*(2), 1339–1340. https://doi.org/10.1021/acs.est.0c07842

Weisberg, D. S., Hirsh-Pasek, K., & Golinkoff, R. M. (2013). Guided play: Where curricular goals meet a playful pedagogy. *Mind, Brain, and Education, 7*(2), 104–112. https://doi.org/10.1111/mbe.12015

Printed in the United States
by Baker & Taylor Publisher Services